W9-AWH-432

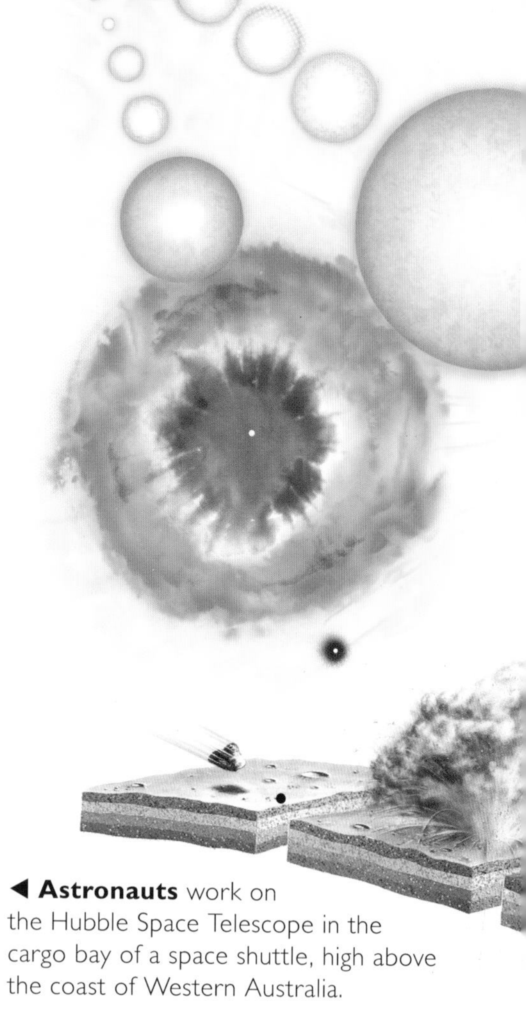

◀ **Astronauts** work on the Hubble Space Telescope in the cargo bay of a space shuttle, high above the coast of Western Australia.

◁ **Overleaf** Huge clouds of gas and dust in the Eagle Nebula where stars are forming.

space explained

A Beginner's Guide to the Universe

ROBIN SCAGELL

A Henry Holt Reference Book
Henry Holt and Company
New York

A Henry Holt Reference Book
Henry Holt and Company, Inc.
Publishers since 1866
115 West 18th Street
New York, New York 10011

Henry Holt® is a registered trademark
of Henry Holt and Company, Inc.

Published in Canada by Fitzhenry & Whiteside Ltd.,
195 Allstate Parkway, Markham, Ontario L3R 4T8

Library of Congress Cataloging-in-Publication Data
Scagell, Robin.
Space explained: a beginner's guide to the universe / Robin
Scagell.—1st ed.
p. cm.—(A Henry Holt reference book)
Includes bibliographical references and index.
1. Astronomy—Amateurs' manuals. I. Title. II. Series
QB64.S42 1996 96-12122
520—dc20 CIP

ISBN 0-8050-4872-3

First American Edition—1996

This book was conceived, edited, and designed by
Marshall Editions
170 Piccadilly, London W1V 9DD

Editor: Katrina Maitland Smith
Designers: Ralph Pitchford, Steve Woosnam–Savage
Managing Editor: Kate Phelps
Art Director: Branka Surla
Editorial Director: Cynthia O'Brien
Production: Janice Storr, Selby Sinton
Research: Lynda Wargen
Consultants: Dr Peter J. Andrews, Royal Greenwich Observatory, Cambridge.
Duncan L. Copp, Mill Hill Observatory, University College, London.

Printed and bound in Portugal by Printer Portugesa
Originated in Singapore by Master Image
All first editions are printed on
acid-free paper.∞

10 9 8 7 6 5 4 3 2 1

Contents

WHAT IS SPACE?

THE SOLAR SYSTEM

WHAT IS SPACE?

SPACE IS EVERYTHING THAT LIES OUTSIDE THE EARTH, THE PLANET ON WHICH WE live. In space, there are many different objects, such as other planets, comets, black holes, and stars. Our sun is just a medium-sized star in a galaxy of one hundred billion stars. Outside our galaxy are billions of other galaxies.

Looking into space

The study of objects in the night sky is called astronomy. Although humans have traveled no farther than the Moon, we know a lot about the other objects we can see in the sky.

The brightest objects are also some of the closest. This book starts with the brightest—the Sun—and works its way outward to the faintest, which are at the very edge of space.

▼ **As we look up at the sky,** the brightest objects are some of the closest. We can see the nearest galaxies using just our eyes, but binoculars or telescopes are needed for the fainter and more distant objects. As we move outward, the distances between objects get larger.

observatory on mountain

radio telescope

aurora and meteors

artificial satellite

the planets

▼ When we talk about days and years, we are measuring time by the spin of the Earth and its orbit around the Sun. Each day, the Sun appears to move around our sky (*see right*), but it is not the Sun that is moving—the effect is caused by the spin of the Earth as it orbits the Sun.

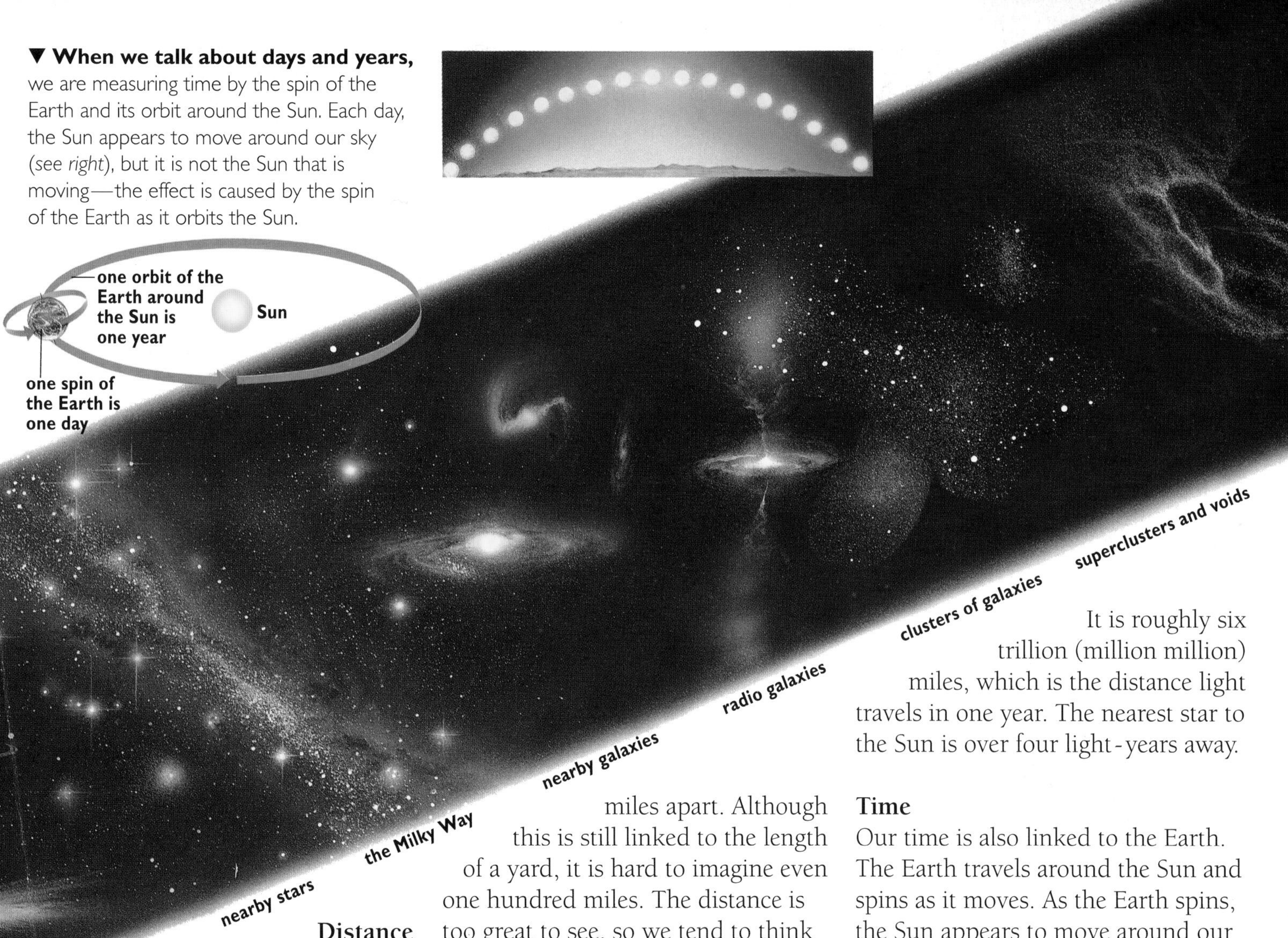

Distance

The huge distances in space are hard to imagine, because they are unlike anything we are used to. But all the distances are worked out step by step by linking them to what we know on Earth. It is the way we measure any distance. We know how long an inch is, and a foot, and we can judge what a yard looks like because we know that it is three feet long.

A mile is 1,760 yards long. Cities may be hundreds or thousands of miles apart. Although this is still linked to the length of a yard, it is hard to imagine even one hundred miles. The distance is too great to see, so we tend to think of one hundred miles as maybe a couple of hours journey by car.

It is the same in astronomy. Distances and sizes of planets can be measured in miles, but we cannot imagine one million miles any better than one hundred miles. So, we think of journey times again, but now we use the travel time of light, which is 186,300 miles per second.

Distances to stars and other galaxies are measured in light-years. A light-year is a distance, not a time. It is roughly six trillion (million million) miles, which is the distance light travels in one year. The nearest star to the Sun is over four light-years away.

Time

Our time is also linked to the Earth. The Earth travels around the Sun and spins as it moves. As the Earth spins, the Sun appears to move around our sky. The time it takes for the Earth to spin once—and for the Sun to return to the same position in the sky—is one day.

It takes a year for the Earth to travel once around the Sun. There are 365.25 days in a year—that is, the Earth spins 365.25 times in one trip around the Sun. Because there is not a complete number of days in a year, we make our year 365 days long and add one day to make it 366 days long every four years.

How things work

▲ **A space shuttle needs** powerful rockets to give it enough speed to ,orbit the Earth without falling back toward the planet's surface.

The Earth's path around the Sun is called its orbit. Almost everything in space is held in orbit around something else, and the force that makes this happen is gravity. This is the same force that makes objects fall to the ground on Earth and that keeps us on Earth's surface. As you bicycle up a hill, you have to work against gravity. Freewheel down the other side and gravity is helping by pulling you toward the center of the Earth.

The force of gravity pulls every object toward every other object. But the force is tiny unless one of the objects is massive. That is why we feel the gravity from the Earth and not from our friends or from houses.

Beating gravity

For a spacecraft to leave Earth's surface and go into space, it must beat the force of gravity, which is pulling it back. If you throw a ball into the air, it falls back. Throw it harder and it goes higher. If you could throw it at 17,700 mph, it would never fall back to Earth. At this speed, a spacecraft will beat the pull of Earth's gravity and will go into orbit around the planet. Although the spacecraft still feels the pull of Earth's gravity, its speed around the planet balances the pull.

Mass, gravity, and weight

The amount of material in an object is called its mass. On Earth's surface, the pull of gravity on that mass is the object's weight. In space, objects do not have weight, since they are not on Earth's surface, but they still have mass. And the more mass an object has, the greater will be the pull of its gravity on anything around it.

in orbit around Earth

pulled back to Earth by gravity

escape velocity

◀ **Anything moving upward** is pulled back by the Earth's gravity. If it is moving too slowly, it falls back to the ground, while at a speed of 17,700 mph it goes into orbit around the Earth. At 25,000 mph (called the escape velocity), it leaves Earth altogether.

The central mass heats up to become the Sun (*right*). The largest lumps in the disk join to form planets, which are battered by the smaller lumps that are left (*far right*).

▶ **The Sun and planets** formed from gas and dust drawn together by gravity (*from top*). Most of the material was at the center and became the Sun, while a disk of material spinning around it gathered into clumps, which became the planets.

The Sun formed, probably with other new stars, in a swirling cloud of gas, ice, and dust.

As the sun begins to form, material in the cloud clumps together to form a central mass surrounded by a disk (*above left*), which contains lumps of gas, ice, and dust (*above*).

The solar system is born, with all the planets moving in the same direction around the Sun (*left*). Volcanoes on the rocky planets belch gases, creating atmospheres (*below*).

A Simple Guide to Everything

Everything is made of tiny particles called protons, neutrons, and electrons, which are arranged in groups called atoms. The different types of atoms are known as elements. Hydrogen is the simplest and lightest element, with just one proton and one electron. The more protons an element has, the heavier it is. Carbon (see right), which is in all living things, has six each of protons, neutrons, and electrons.

All material can exist as a solid, a liquid, or a gas. Which state it is in depends on its temperature. For example, as water is heated, it changes from solid ice, to liquid, then to gas. Everything else can exist in these three states. So oxygen, which we usually know as a gas, can be a liquid or a solid if it is cold enough.

The particles that make up matter are constantly moving. The higher the temperature, the faster they move. The particles of a gas move around at high speeds and press on anything in their way. This is what we call pressure.

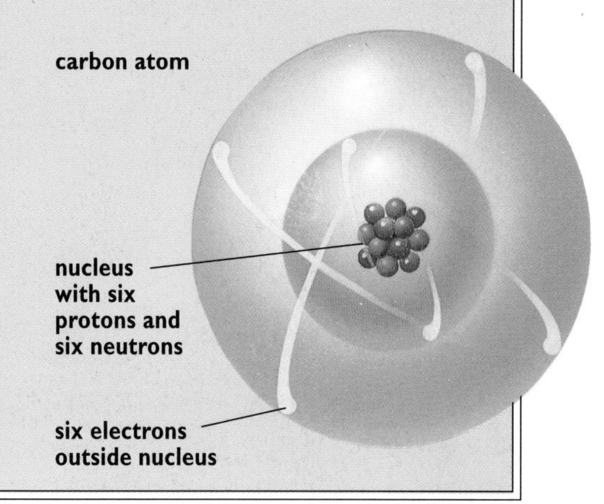

The birth of the solar system

Gravity was the main force that caused the solar system to form. The Sun and planets were born out of a swirling disk of gas, ice, and dust about 4.6 billion years ago. At first, the material throughout the disk was the same—mostly the light elements hydrogen and helium, ice, and dust grains containing heavier elements.

As the Sun, our nearest star, formed (*see pages 38–39*), the gas, ice, and dust around it was drawn into clumps by the pull of gravity. Close to the Sun, the clumps were mostly made of heavier, rocky materials because the lighter ones were blown farther out by the Sun's heat. In the outer part of the solar system, the lighter elements could clump together.

The clumps themselves joined together, forming bigger and bigger clumps, until they became planet-sized. Material that was left hit the surfaces of the new planets and moons, leaving craters.

At the edge of the rocky zone of the solar system, between Mars and Jupiter, there was not enough material to make a planet. The thousands of small, rocky bodies that remain make up what is known as the asteroid belt. Farther out, a similar, colder, region of icy clumps became the Kuiper Belt.

THE SOLAR SYSTEM

SOME FOUR AND A HALF BILLION YEARS AGO, OUR SUN AND THE NINE PLANETS that orbit around it were formed together in space. This is our solar system. Countless smaller objects, such as asteroids and comets, are also part of this system, which is held together by the Sun itself.

Because the Sun is so bright, you should never look straight at it. Its light can seriously damage your eyes and even blind you. Astronomers use special methods to study the Sun safely.

Our sun

For life on Earth, the Sun is the most important thing there is. It produces almost all the energy we have—even the energy inside coal and oil originally came from the Sun. Because the Sun is the largest body in the solar system, all the planets and other bodies are held in orbit around it by the pull of the Sun's gravity.

The Sun is Earth's nearest star. If you were to travel far into space, the Sun would look just like the other stars. And, like the stars, it is a huge globe of hydrogen and helium gas which produces enormous amounts of heat and light. How do the Sun and stars do this?

▲ **A total solar eclipse** happens when the Moon is directly in front of the Sun. For a few minutes, the Moon covers the Sun, and we can see the Sun's outer atmosphere, called the corona, around the disk of the Moon.

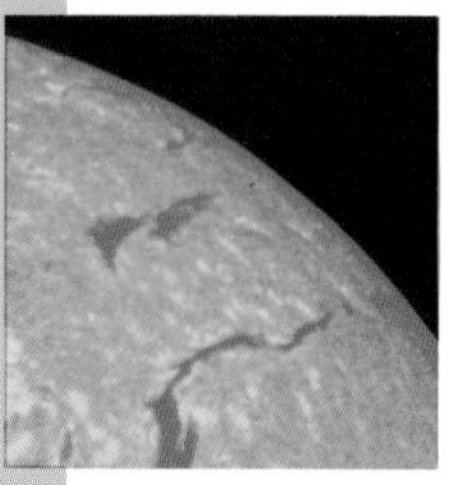

hydrogen

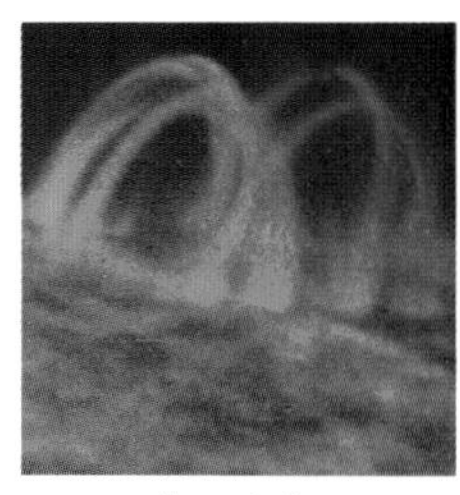

ultraviolet

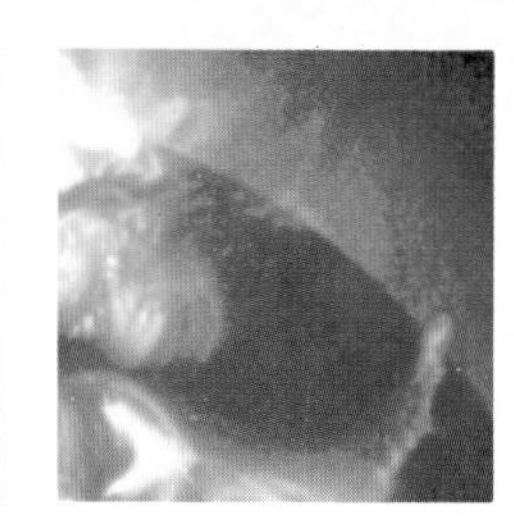

X-ray

▲ **Special equipment shows** different types of radiation coming from above the Sun's surface. Sunspots can be clearly seen in the red light of hydrogen from just above the surface. Higher up, ultraviolet light shows prominences (see *also opposite*), while X-rays show activity in the Sun's outer atmosphere.

▶ **Sunspots look dark** because they are four times fainter than the rest of the Sun. If you could see them by themselves, they would be brilliant. The layers of the Sun's atmosphere—the chromosphere and the corona—get hotter as you go outward.

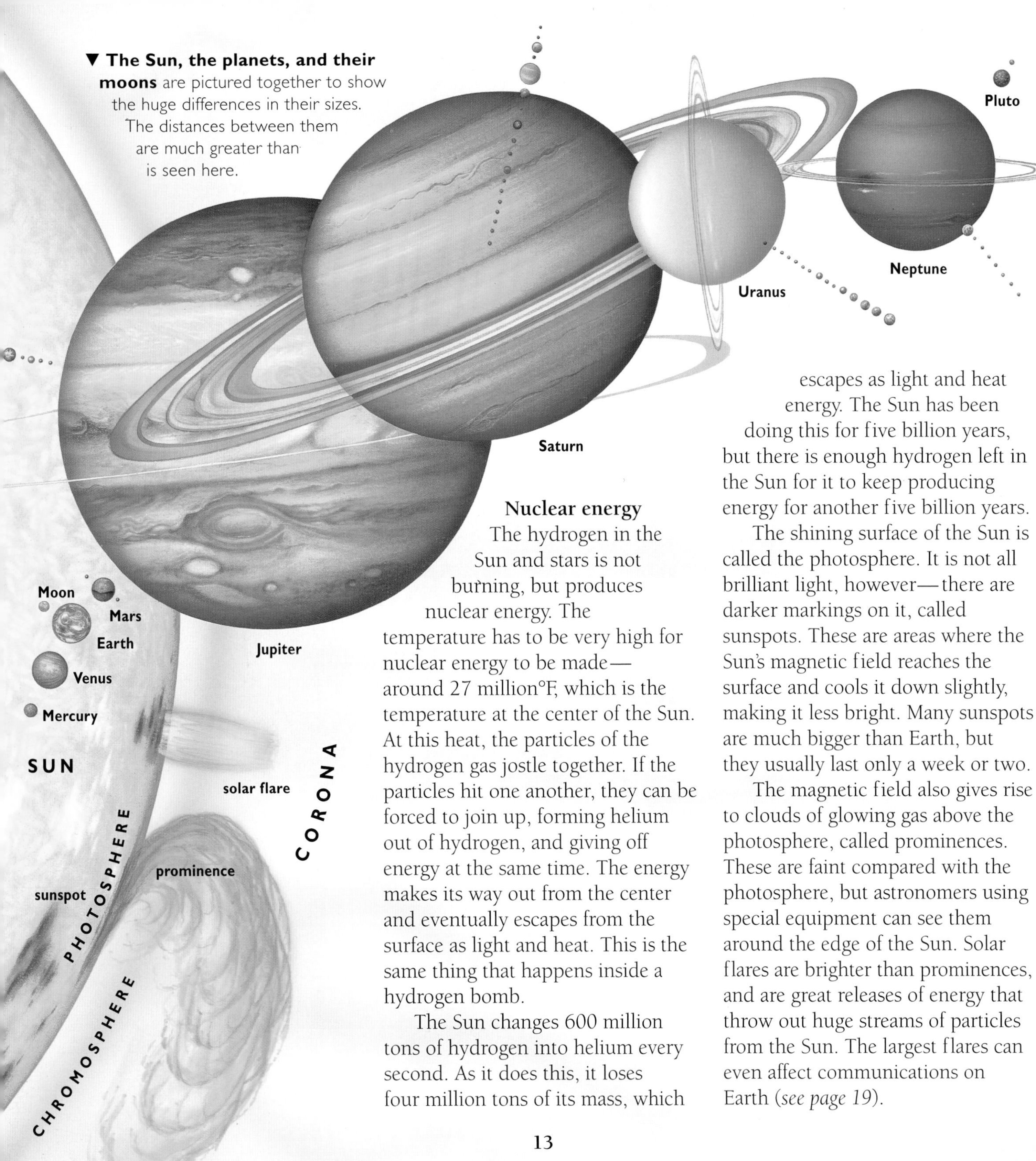

▼ **The Sun, the planets, and their moons** are pictured together to show the huge differences in their sizes. The distances between them are much greater than is seen here.

Nuclear energy

The hydrogen in the Sun and stars is not burning, but produces nuclear energy. The temperature has to be very high for nuclear energy to be made—around 27 million°F, which is the temperature at the center of the Sun. At this heat, the particles of the hydrogen gas jostle together. If the particles hit one another, they can be forced to join up, forming helium out of hydrogen, and giving off energy at the same time. The energy makes its way out from the center and eventually escapes from the surface as light and heat. This is the same thing that happens inside a hydrogen bomb.

The Sun changes 600 million tons of hydrogen into helium every second. As it does this, it loses four million tons of its mass, which escapes as light and heat energy. The Sun has been doing this for five billion years, but there is enough hydrogen left in the Sun for it to keep producing energy for another five billion years.

The shining surface of the Sun is called the photosphere. It is not all brilliant light, however—there are darker markings on it, called sunspots. These are areas where the Sun's magnetic field reaches the surface and cools it down slightly, making it less bright. Many sunspots are much bigger than Earth, but they usually last only a week or two.

The magnetic field also gives rise to clouds of glowing gas above the photosphere, called prominences. These are faint compared with the photosphere, but astronomers using special equipment can see them around the edge of the Sun. Solar flares are brighter than prominences, and are great releases of energy that throw out huge streams of particles from the Sun. The largest flares can even affect communications on Earth (*see page 19*).

The Moon in the sky

Everyone knows what the Moon looks like. We know that it sometimes looks crescent shaped and sometimes round, but often without understanding how or why it changes. When and where will we see it as a crescent New Moon? When is it at Full Moon? Why is it bigger when it is rising?

The Moon is Earth's natural satellite. A satellite is an object that is held in orbit around a larger object, such as a planet. We call the Moon a "natural" satellite because it is not man-made. It is held in its orbit by the pull of Earth's gravity.

We only see the Moon because the Sun shines on it. The Moon does not produce its own light. As the Moon orbits the Earth, its position in the sky changes, so the direction of the Sun's light on it also changes. This creates the different shapes of the sunlit part of the Moon as it moves from New Moon to Full Moon and back again. They are known as the phases of the Moon. There are 29.5 days—just about a month—between two New Moons. By following the Moon's progress over a month, we can find out more about its phases.

▲ **The Moon** looks much bigger when it is low in the sky, but this is just an illusion because we are comparing it with objects on the horizon. When it is higher, there is nothing familiar to compare it with.

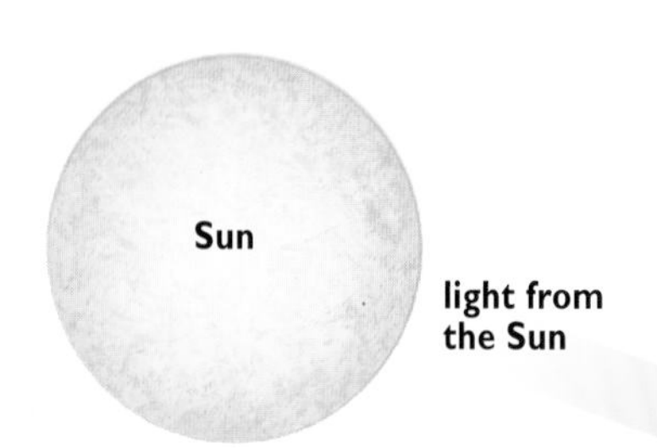

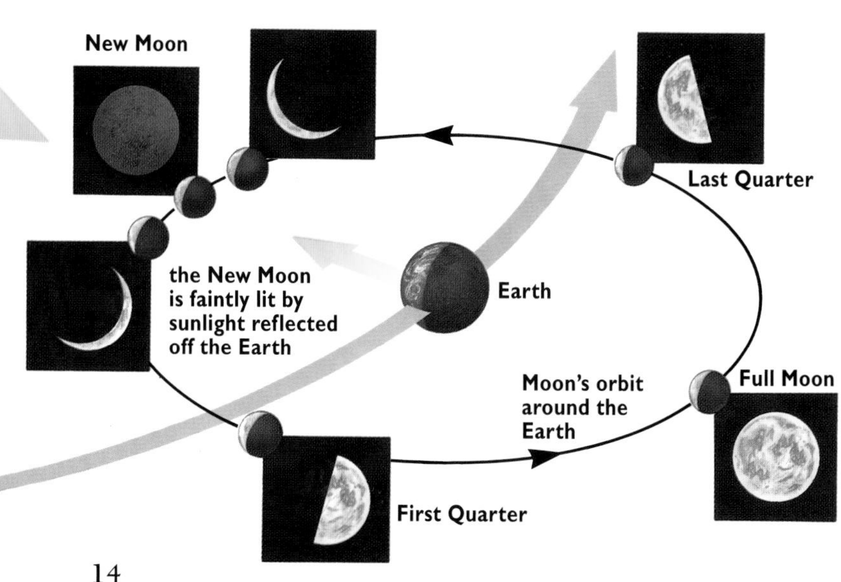

▶ **How much we can see** from Earth of the bright side of the Moon depends on where the Moon is in its orbit. The changing shapes of the bright part of the Moon are known as the Moon's phases. The Moon turns once in exactly the same time that it takes to orbit the Earth, so it always keeps the same side toward us.

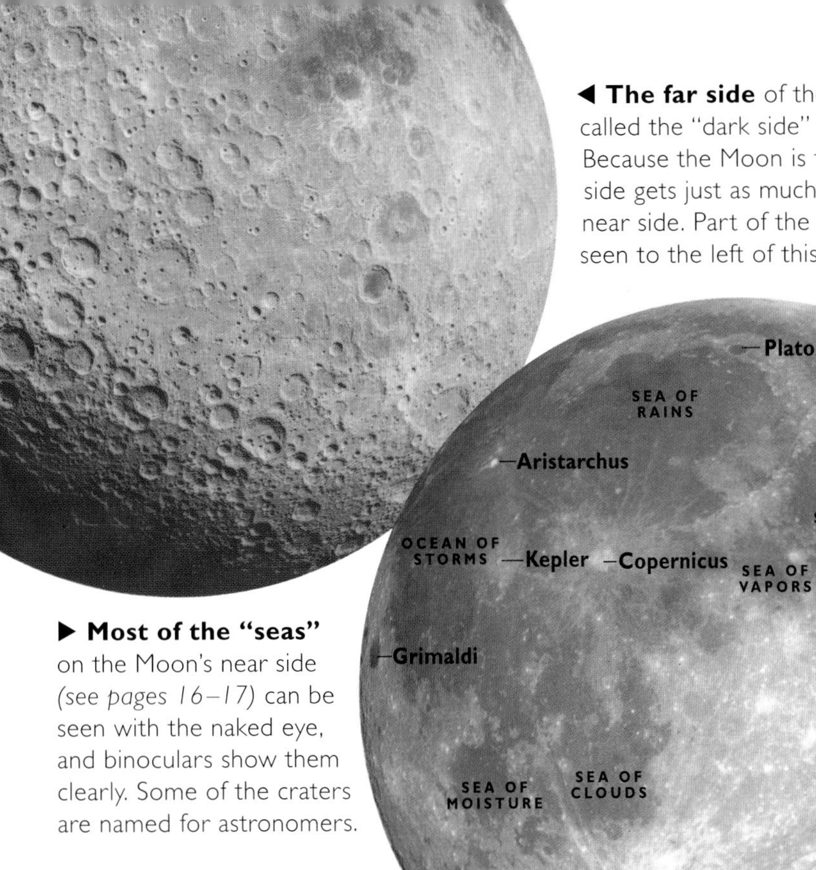

◀ **The far side** of the Moon is often called the "dark side" but this is wrong. Because the Moon is turning, the far side gets just as much sunlight as the near side. Part of the near side can be seen to the left of this photograph.

MOON DATAFILE

Distance from Earth 238,900 miles
Diameter 2,160 miles (0.27 Earths)
Mass 0.012 Earths
Day (time taken to spin once) 27 Earth days 8 hours
Time taken to orbit the Earth 27 Earth days 8 hours
Surface temperature 248°F (day) –292°F (night)
Maximum magnitude (see *page* 45) –12.7

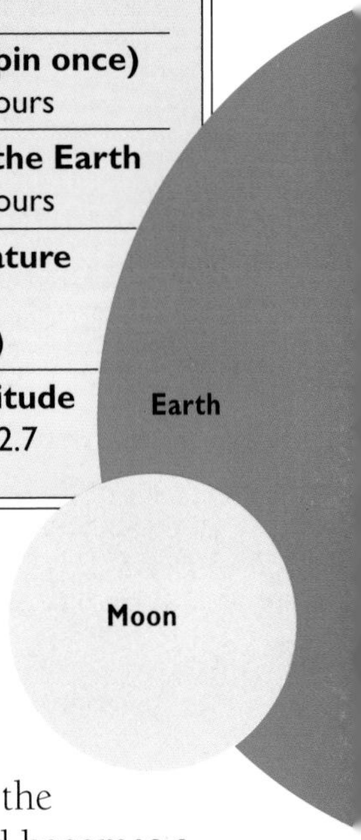

Earth

Moon

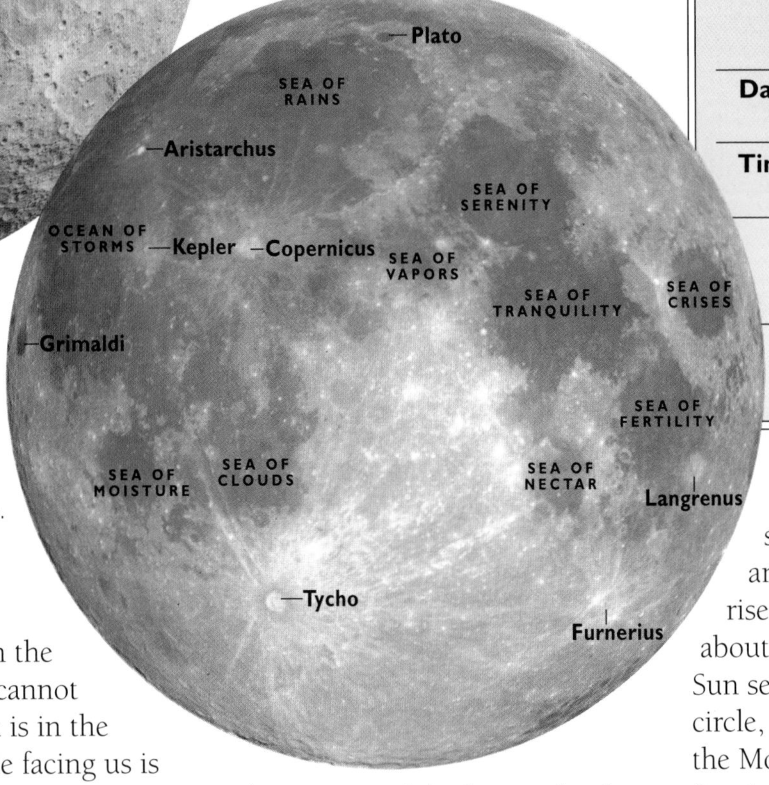

Plato
SEA OF RAINS
Aristarchus
SEA OF SERENITY
OCEAN OF STORMS
Kepler
Copernicus
SEA OF VAPORS
SEA OF TRANQUILITY
SEA OF CRISES
Grimaldi
SEA OF FERTILITY
SEA OF MOISTURE
SEA OF CLOUDS
SEA OF NECTAR
Langrenus
Tycho
Furnerius

▶ **Most of the "seas"** on the Moon's near side *(see pages 16–17)* can be seen with the naked eye, and binoculars show them clearly. Some of the craters are named for astronomers.

The Moon's phases

Sometimes the Moon is almost in a line between the Earth and the Sun. We cannot see the Moon because it is in the daytime sky and the side facing us is not lit up. This is the true New Moon. A couple of days later, it has moved to one side of the Sun, and we see it as a thin crescent, which people generally call the New Moon.

You will see a crescent New Moon only in the western sky soon after sunset. The crescent's horns always point upward because the Sun, which is much farther away, is shining on the Moon from behind and below it.

Four or five days later, the Moon has moved around in its orbit and can be seen much higher in the sky, later in the evening. Now it is a half-circle. In fact, this phase is called the First Quarter, because the Moon is a quarter of the way around its orbit. This is a good time to look at the Moon with binoculars. You will see the surface features clearly because they cast long shadows in the Sun's light which is coming from the side.

As the Moon moves farther around its orbit, it rises later and later in the evening. The half-circle seems to swell as we see more of the sunlit side. Halfway around its orbit, it rises in the east at about the same time as the Sun sets in the west, and becomes a circle, called the Full Moon. Because the Moon is directly opposite the Sun in the sky, no shadows can be seen on its surface. The same thing happens on Earth—when the Sun is directly above us, we do not cast a shadow on the ground around us.

From now on, the Moon goes back through its phases, but with the light coming from the other side. When it is a half-circle again, it is said to be in its Last Quarter because it is three quarters of the way around its orbit. It rises later and later, until it is a thin crescent again, rising an hour or so before the Sun comes up.

The Moon in close-up

Look at the Moon through a telescope, and a new world springs into view. There are big differences between the Earth and the Moon. The Moon has no air, no water, and no life. The landscape is not worn away by wind and rain, so it will stay the same for millions of years.

There are not even any large, active volcanoes to create new landscapes, because the Moon has a more solid interior than the Earth. But there are signs that the surface was once churned up. There are huge craters and hardened flows of lava.

▲ **The Moon's dark areas,** such as the crater Tsiolkovskii, are made of volcanic lava called basalt. Bubbles in a basalt sample (*above left*), which was brought back by astronauts from Apollo 15, show that the rock was heated at some time.

▼ **Almost all the craters** on the Moon were formed by collisions. Small craters are simply bowl shaped, but those more than about 20 miles across often have flatter floors, sometimes with a central peak.

An impact crater is created when a large object hits the Moon's surface.

The fierce heat created by the impact completely destroys the object in a giant explosion.

A crater is left that is many times the size of the object that made it. Material may rebound into the middle, making a central peak.

Large impacts can create lava flows, which may fill older craters.

Craters and "seas"

Some of the Moon's craters are more than 100 miles across—if you were standing in the center, you would not be able to see the crater walls. Even a city like New York or Chicago would fit inside a crater with plenty of room to spare. Most of the craters formed more than three billion years ago when large chunks of material left over from the formation of the solar system (*see page 11*) crashed onto the surface.

The collisions released molten lava from inside the Moon. It flowed across the surface before cooling and setting hard, creating the Moon's flatter, darker areas. The astronomers who first looked at the Moon through telescopes a few hundred years ago called these areas "seas" or, in Latin, *maria*. The romantic names they gave to the seas, such as "Sea of Tranquility," are still used today.

Lava flows have filled some of the older craters, leaving just their rims sticking up above the surface as "ghost craters." Since the seas were formed, there have been fewer collisions, so the seas are less marked. Some of the more recent collisions threw material a long way, leaving pale rays around the craters. These craters, such as one called Tycho, are known as ray craters.

Men on the Moon

Project Apollo landed 12 men on the Moon between 1969 and 1972. It was a major space program put together by the United States.

There were three main parts to the Apollo spacecraft. The Command Module was the main living quarters for the three crew during their journey, while the Service Module contained supplies such as fuel and oxygen. The landing craft, called the Lunar Excursion Module, was used only for the descent from the orbit of the Moon to the Moon's surface and back again. Two men went to the surface, spending up to three days there. During this time, the third astronaut stayed in the Command Module in orbit around the Moon.

The astronauts who walked on the Moon were able to cover a greater distance than usual with each stride. This is because the Moon is smaller than the Earth and is made of lighter materials, so the pull of its gravity is only one-sixth that of Earth. This means that the force of each step on the Moon takes an astronaut farther. On the Moon, your weight would be only one-sixth of what it is on Earth.

The Apollo astronauts set up experiments and brought back samples of rock to help us find out more about the Moon. Since 1972, however, no astronaut has traveled outside Earth's orbit. A major reason for this is the huge cost of manned missions into space.

▼ **There were 17 Apollo launches,** but it was only the last seven that set out to land their crews on the Moon. Each mission took two and a half days to reach the Moon, but Apollo 13 did not land because of an explosion in its oxygen tank.

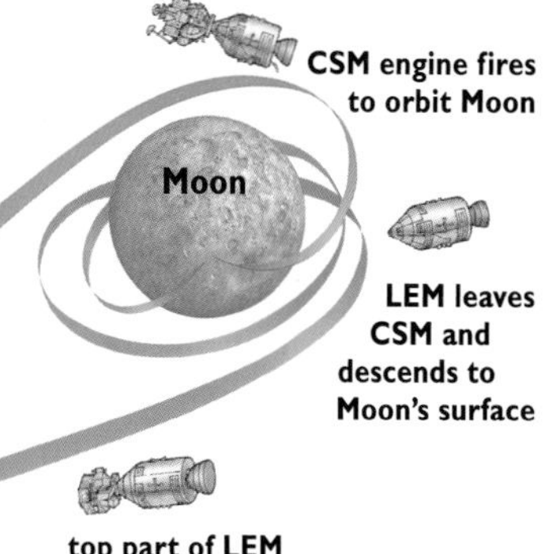

escape rocket
Command Module
Service Module
Lunar Excursion Module
third stage
second stage
first stage
splash-down in Pacific Ocean
Command Module separates from Service Module before entering Earth's atmosphere
Earth
launch from Cape Kennedy, U.S.A.
CSM turns around and pulls LEM from third stage
third stage contains Lunar Excursion Module (LEM), with Command and Service Modules (CSM) on top

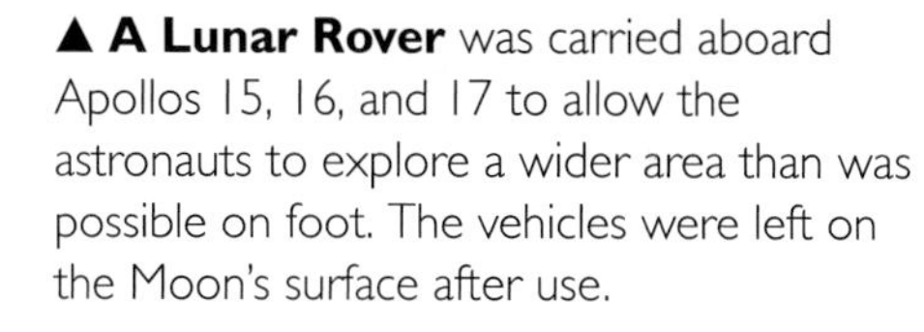

▲ **A Lunar Rover** was carried aboard Apollos 15, 16, and 17 to allow the astronauts to explore a wider area than was possible on foot. The vehicles were left on the Moon's surface after use.

▶ **The Saturn 5 rocket** used for the Apollo missions (pictured here with cutaways to show inside) was the most powerful rocket ever built. It had three stages. The lower two dropped away during the launch, but the third went into Earth's orbit carrying the Command, Service, and Lunar Excursion Modules.

▲ **The beautiful aurora** seen in the north is called the Northern Lights (the scientific name is *aurora borealis*). The aurora in the south is called the Southern Lights (*aurora australis*).

The Earth's atmosphere

We live at the bottom of an atmosphere which provides us with air to breathe and shields us from radiation from space. At the same time, it is responsible for spectacular displays which originate in space—the colorful auroras and the fiery trails of meteors.

The Earth's atmosphere is made mostly of nitrogen, a gas that is harmless to us. About a fifth of it is oxygen, which is what we breathe to keep us alive. There are also small amounts of other gases, such as carbon dioxide, the gas we breathe out.

As you go higher, the atmosphere thins out—there is less of it, so it is harder to breathe. Eventually, about 60 miles up, it is so thin that you are almost in space. But even this high up, the atmosphere cannot be ignored. Spacecraft that orbit the Earth travel above about 125 miles, and the pull of the atmosphere is still enough to make them lose height, although it happens gradually, over months or years.

▼ **Meteors are often called shooting stars,** but they are tiny pieces of dust from comets. As they race through the upper atmosphere, they burn up, leaving a fiery trail.

▶ **Each aurora forms a ring** of about 2,500 miles around one of the Earth's magnetic poles. The thicker air lower down in the atmosphere glows green, and the upper air glows red. Sometimes there are bright pulses, as if a lamp is being switched on and off.

The aurora

Around the Earth's north and south poles, the night skies are often lit up by the auroras. They look like giant, colored curtains hanging in the sky, moving slowly as you watch.

The auroras are caused by the solar wind, which is made up mostly of protons and electrons given off by the Sun. Protons and electrons are affected by magnetism. Unlike gravity, which affects everything, magnetism only affects certain things, such as some metals.

The area around a magnet where the pull of the magnet is felt is called its magnetic field. The Earth has a magnetic field, probably caused by iron deep inside it. This field is strongest toward the Earth's north and south poles, in regions called the magnetic poles.

The particles of the solar wind passing Earth are attracted to the magnetic poles. They collide with Earth's upper atmosphere at speeds of up to 1,250 miles a second, making the gases in the thin air glow. Streams of these particles cause the beautiful auroras.

Sometimes, solar flares create more of a solar gale than a solar wind (*see page 13*), and the aurora can be seen over a much wider area. These larger streams of particles can cause magnetic storms on Earth, which affect radio waves and cause bad communications.

◀ **Very rarely,** large rocks from space make craters on Earth's surface, such as this one at Wolf Creek in Australia. The rock was destroyed in the collision. Smaller rocks survive because they are slowed down by the Earth's atmosphere before they hit. These are called meteorites.

Meteors

Larger particles collide with Earth's atmosphere, too. Our part of the solar system is a dusty place, with chunks of material flying about in all directions. Most of the chunks are as tiny as grains of instant coffee, and move at about 25 miles per second.

When one of these grains runs into the Earth's atmosphere at this speed, it heats up. In less than a second, the grain is so hot that it starts to burn up, causing a fiery trail in the sky some 50 miles above us. It looks as if a star has come loose and fallen from the sky, which gives it the popular name of shooting star, though the correct name is meteor.

You can usually see a few meteors an hour if you watch the sky away from city lights. Sometimes the Earth passes through streams of these particles, which are the remains of comets. When this happens, you may be able to see as many as one every minute.

The Earth's Magnetic Field

As the solar wind passes the Earth, it is affected by the planet's magnetic field. The particles in the solar wind are drawn down toward the Earth's magnetic poles (see diagram at right), *where they hit the upper atmosphere and cause the auroras. The solar wind in turn affects the magnetic field, which is drawn into a long tail on the side of Earth that faces away from the Sun.*

north magnetic pole
solar wind
Earth
magnetic field
south magnetic pole

▲ **The head of a comet** is quite fragile. Before Comet Shoemaker-Levy 9 crashed into the planet Jupiter (see *pages 26–27*), it had been broken up into more than 20 pieces by the pull of Jupiter's gravity.

Comets

Comets excite more interest than almost any other kind of object in the sky. A bright comet is a rare event. It can appear without warning and looks like a ghostly knife hanging in the sky, with a bright star at its point. It moves slowly from night to night until it has passed out of sight, perhaps not to be seen again for thousands of years. People in the past thought that these strange objects were signs that something bad might happen.

Although we now know what comets are, they can still appear when we do not expect them. Even when we know one is on its way, it can be fainter or brighter than we thought it would be. It may even break up and disappear altogether.

tails shrink as comet moves away from Sun

comets can approach the Sun from any direction

solar wind

Earth's orbit

Sun

Earth

tails grow as comet nears Sun

comet's orbit

▼ **The two tails** of a comet show up in this photograph of Comet West. The faint, blue gas tail can be clearly seen below the whiter dust tail. A comet's dust tail can be millions of miles long and shines in the sunlight.

Frozen beginnings

Comets come from the farthest edges of the solar system. Well beyond the orbit of the planet Pluto lies material left over from the birth of the solar system itself. It consists of chunks of ice and dust that have been in deep freeze for billions of years.

There are probably millions of these chunks, of all sizes, from tiny splinters to great blocks hundreds of miles across. Most of them stay in lonely orbits around the Sun, held there by the Sun's gravity. Now and then, one of them may be nudged very slightly from its orbit, perhaps by the effects of the gravity of a nearby star. The comet starts to move toward the inner solar system, speeding up as it goes.

As the comet nears the Sun, the outer layers of the ice warm up and turn to gas, which forms a haze around the comet's central body (the nucleus). When the comet gets inside the orbit of Jupiter, the haze is affected by the solar wind (*see page 19*), which blows the gas into a long tail. Any dust that was frozen in the outer layers of the nucleus also streams away, but as a separate tail.

The comet's orbit swings it around the Sun, and it then begins to move back out into deep space,

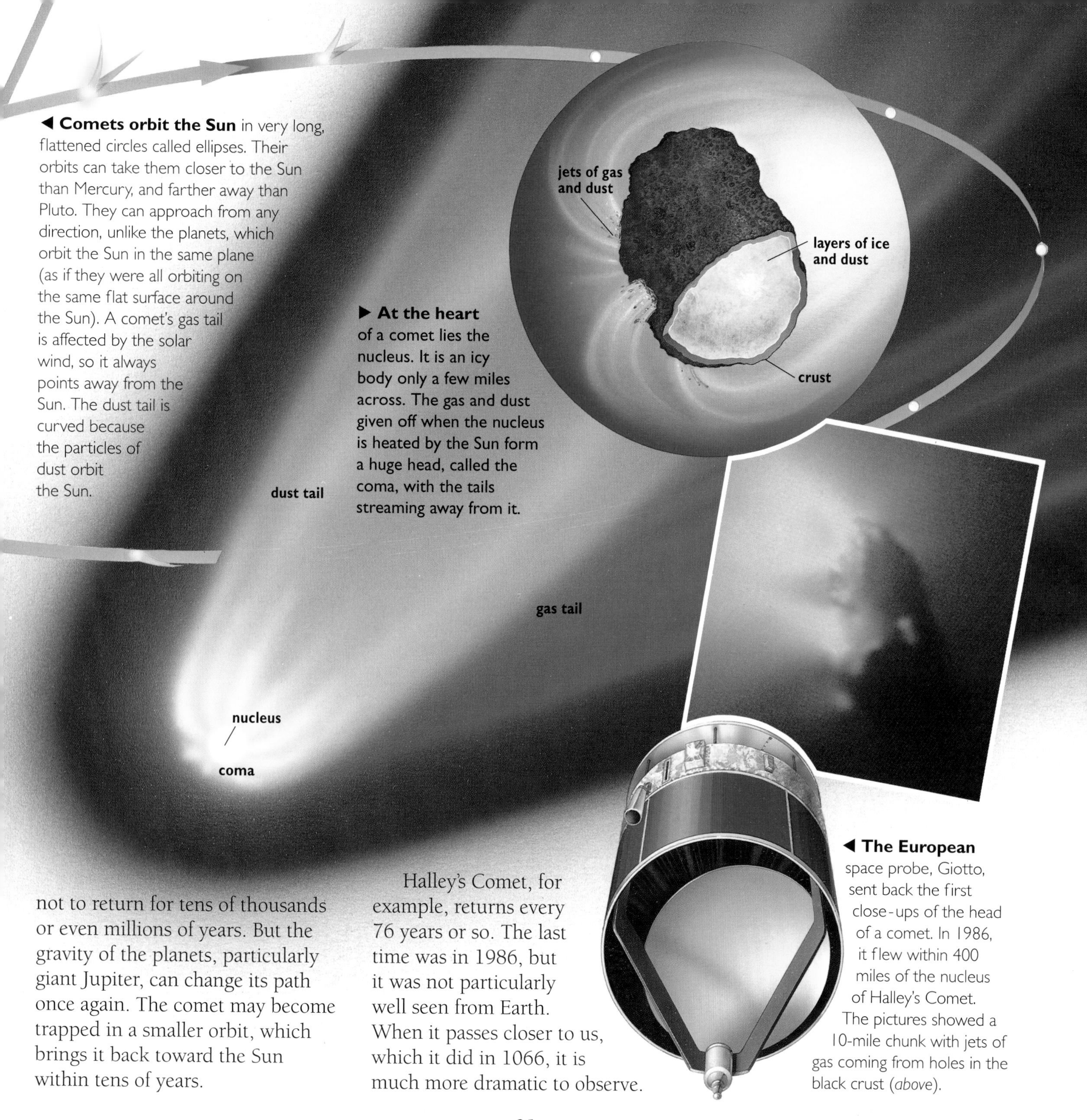

◀ **Comets orbit the Sun** in very long, flattened circles called ellipses. Their orbits can take them closer to the Sun than Mercury, and farther away than Pluto. They can approach from any direction, unlike the planets, which orbit the Sun in the same plane (as if they were all orbiting on the same flat surface around the Sun). A comet's gas tail is affected by the solar wind, so it always points away from the Sun. The dust tail is curved because the particles of dust orbit the Sun.

▶ **At the heart** of a comet lies the nucleus. It is an icy body only a few miles across. The gas and dust given off when the nucleus is heated by the Sun form a huge head, called the coma, with the tails streaming away from it.

◀ **The European** space probe, Giotto, sent back the first close-ups of the head of a comet. In 1986, it flew within 400 miles of the nucleus of Halley's Comet. The pictures showed a 10-mile chunk with jets of gas coming from holes in the black crust (*above*).

not to return for tens of thousands or even millions of years. But the gravity of the planets, particularly giant Jupiter, can change its path once again. The comet may become trapped in a smaller orbit, which brings it back toward the Sun within tens of years.

Halley's Comet, for example, returns every 76 years or so. The last time was in 1986, but it was not particularly well seen from Earth. When it passes closer to us, which it did in 1066, it is much more dramatic to observe.

Mercury and Venus

No two planets in the solar system are the same. The conditions on each one depend mostly on its size and distance from the Sun.

The bigger the planet, the greater the pull of its gravity will be on any gas that surrounds it. This gas makes up its atmosphere, and without an atmosphere a planet is a dry, rocky world with no clouds, rain, or oceans. This is because water can only exist as a liquid if there is enough pressure from the atmosphere above it to stop it from turning straight into a gas.

The closer a planet is to the Sun, the hotter it will be. This will also affect whether there is an atmosphere. If the planet is very close to the Sun, any atmosphere will simply boil away.

Mercury and Venus show us how a planet's size and distance from the Sun can make a big difference in what the planet is like.

Venus
Mercury
Sun

▲ **Mercury and Venus** stay quite close to the Sun in our sky because their orbits are inside that of the Earth. They can be seen only in twilight skies, and Mercury never appears high in the sky when the Sun is below the horizon. In this diagram, the planets have been drawn larger than you will see them, so that their phases show more clearly.

◀ **The impact that** created Mercury's Caloris Basin was so great that it nearly broke up the planet. The shock waves went right through the planet and created a strange landscape of jumbled rocks on the other side.

◀ **Mercury's cratered surface** is baked hard by the Sun—the temperature on the planet can be more than 800°F by day, but drops to about –275°F by night.

Swift Mercury

Mercury has always been linked with swiftness. It appears for a little while low in our evening sky just after sunset. Then, only five weeks later, it is in the morning sky before sunrise. It always stays close to the Sun.

The reason for this behavior is that Mercury is the closest planet to the Sun, so it orbits the Sun quickly. Mercury takes 88 days for one orbit, compared with Earth's 365 days.

Because Mercury is a small planet and is close to the Sun, it has almost no atmosphere, and the temperature can reach above 800°F during the day. Like our Moon, it is covered with craters. These are the result of collisions with material from space billions of years ago.

One of these surface features, the Caloris Basin, is huge. It has a central crater more than 800 miles across, surrounded by a bull's-eye of several rings of mountains, covering 2,300 miles. The impact that caused it must have been enormous.

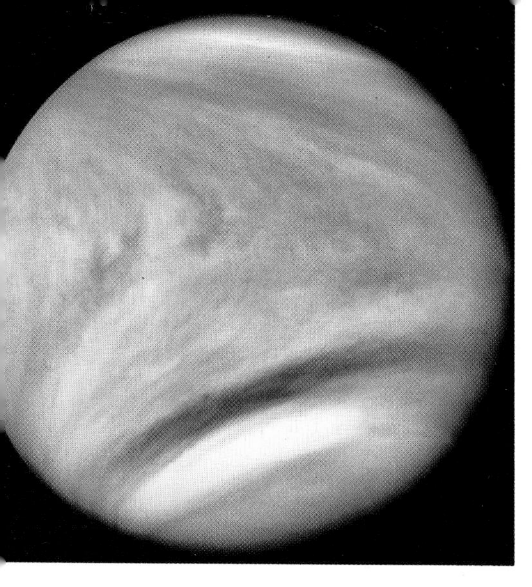

▲ **Venus has few markings to look at,** but ultraviolet light shows cloud patterns which move around the planet in four days.

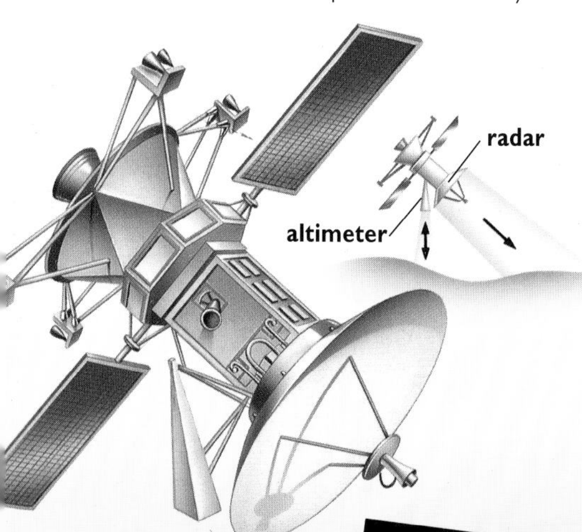

▶ **Pictures of Venus's surface** came from the Magellan spacecraft (*above*), which orbited the planet from 1990 to 1994. It used radar to "see" through the clouds, and an altimeter to measure the height of the surface.

Bright Venus

Venus looks like a brilliant star in the night sky. Like Mercury, it appears in the twilight sky after sunset or before sunrise, but because it is farther out from the Sun, it moves more slowly.

The planet is only slightly smaller than the Earth, and it has a deep atmosphere. The atmosphere is mostly carbon dioxide, which is known as a "greenhouse gas." This is because it traps the Sun's heat, just as a greenhouse gets warm even on a cold but sunny day. In fact, the temperature on Venus is more than 850°F—far too hot for water to stay as a liquid.

The atmosphere is so thick that its pressure at the surface of the planet is 90 times that of Earth's. Probes that have reached the surface have lasted only a matter of hours before they stopped working.

Venus's surface is hidden from our view by thick clouds of sulfuric acid high in the planet's atmosphere. The sulfur probably came from thousands of volcanoes, some of which may still be active.

Mercury and Venus Datafile

Distance from Sun
Mercury: 36 million miles
Venus: 67.247 million miles

Diameter
Mercury: 3,032 miles (0.38 Earths)
Venus: 7,522 miles (0.95 Earths)

Mass
Mercury: 0.055 Earths
Venus: 0.81 Earths

Day (time taken to spin once)
Mercury: 59 Earth days
Venus: 243 Earth days

Year (time taken to orbit the Sun)
Mercury: 88 Earth days
Venus: 225 Earth days

Surface temperature
Mercury: 800°F (day), –275°F (night)
Venus: 870°F

Moons
Mercury: 0
Venus: 0

Maximum magnitude (*see page 45*)
Mercury: 0
Venus: –4.4

Venus
Earth
Mercury

▼ **Beneath the clouds of Venus** the atmosphere is clear, but the clouds make the light dim and orange. Lava flows and volcanoes mark the landscape.

Mars

▲**Clouds high up in Mars's** thin atmosphere can be seen as a band above the horizon in this picture taken by a Viking spacecraft orbiting the planet. The planet's surface is heavily cratered in the southern hemisphere.

In our night sky, Mars shines almost as red as a glowing coal. On the planet itself, the rocks, the dust, and even the sky are reddish.

At first glance, the landscape on Mars looks like a desert on Earth. There are rocks, dunes of dust, and hazy distant mountains. The Sun shines in a sky which is bright, like that of Earth, so that no stars are seen by day. But Mars is a smaller planet, and the pull of its gravity is only one-third as strong as Earth's.

Because Mars is farther from the Sun, its surface is cold—the temperature is well below freezing for most of the time. It has only a thin atmosphere of carbon dioxide, with about one percent of Earth's air pressure. This is unbreathable. Not only is there no oxygen, but the low pressure would bloat your body and turn your blood into gas.

▶ **The dark markings** on Mars change as wind-blown dust moves around. Before spacecraft provided the first close-up pictures of Mars in the 1960s, these changes were thought to be due to the growth of plant life.

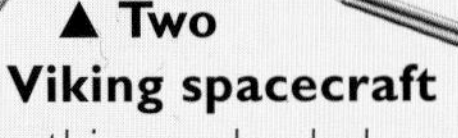

▲ **Two Viking spacecraft** like this one landed on Mars in 1976. They photographed their landing sites (*right*) and tested the soil for signs of life, but found none. The winds blow the orange-red dust around, making the sky look orange.

▼ **The surface of Mars** has different landscapes, some of which are like those on Earth. A huge canyon, called Valles Marineris, crosses one whole side of the planet. It is so large that it could swallow the Rocky Mountains. Mars also has thin clouds, morning mists, and light frosts.

ice clouds

canyon wall

dried-up water channel

crater

On the surface

Mars is a dry planet, with no liquid water. Yet there are signs that this was not always the case. There are many dried-up water channels, and the redness of the rocks was caused by the iron in them combining with a small amount of water.

Scientists believe that, billions of years ago, huge volcanoes threw out gases from the planet's hot center, making a thick atmosphere. Water gushed from beneath the surface, perhaps forming seas. The simplest forms of life, such as algae, may even have existed.

No one knows exactly why it happened, but the planet's atmosphere then leaked away into space, and most of the water was lost. The climate got colder, and most of the water that was left was frozen beneath the surface. There is now only a thin layer of water ice around the planet's poles.

Despite its watery past, Mars is probably completely lifeless today. The volcanoes are now extinct. One of them, Olympus Mons, is the largest volcano in the solar system—it rises 15 miles above the desert that surrounds it.

Martian winds whip up dust storms, sometimes across the whole planet. One field of dust dunes, near the north polar cap, is bigger than Earth's Sahara and Arabian deserts put together.

There are lots of craters, caused by collisions of material from space. Two tiny moons, Phobos and Deimos, dash around the planet—Phobos in 8 hours and Deimos in 28 hours. They are probably chunks of material that, instead of hitting the planet, were caught in orbit around it.

Mars Datafile

Distance from Sun
141.6 million miles

Diameter
4,218 miles (0.53 Earths)

Mass
0.11 Earths

Day (time taken to spin once)
24 Earth hours 37 minutes

Year (time taken to orbit the Sun)
687 Earth days

Surface temperature
70°F (day)
–207°F (night)

Moons
2

Maximum magnitude (*see page 45*)
–2

Earth
Mars

dust storm
thin atmosphere
volcano
mist

Jupiter

▲ **Jupiter (*top*) is one of the brightest** objects in our sky. Through binoculars, you can see its flattened globe and follow the movements of its four larger moons as they orbit the planet.

The largest planet in the solar system, Jupiter is quite unlike any of the planets closer to the Sun.

It is made of the gases hydrogen and helium, and what appears to be its surface is, in fact, the top of its atmosphere, which has swirls of other gases such as methane and ammonia. When a probe from the Galileo spacecraft entered the atmosphere of Jupiter in 1995, it simply kept falling until the increasing heat and atmospheric pressure put it out of action.

▲ **Comet Shoemaker-Levy 9** hit Jupiter in 1994 after it had been broken up into more than 20 pieces. As it entered the atmosphere, each piece was destroyed in an explosion like that of a hydrogen bomb.

Jupiter is 11 times the size of Earth, and its mass is over 300 times greater. Its huge bulk spins once in just under 10 hours. This makes its middle bulge out, and flattens the planet at its poles. Through a telescope, you have to watch for only about 15 minutes before new features come into view as Jupiter turns.

Jupiter's Moons

Callisto

Ganymede

Europa

Io

Future explorers will not be able to visit Jupiter itself, but they can land on its moons. Four of these are very large: Ganymede, Callisto, Europa, and Io.

Ganymede, the biggest moon in the solar system, is larger than Mercury. It is covered with dark and light blotches, probably caused by impacts of material from space that heated and melted the rock in some places. Callisto is also larger than Mercury, and it is completely covered with craters. Europa is smaller and has no craters at all. If you could shrink Europa to the size of a pool ball, it would be just as smooth.

The oddest of all Jupiter's moons is Io, the closest large moon to the planet. Although it is about the same size as our moon, Io is far from being a dead world. There is constant activity on its surface (see below left), with geysers of sulfur shooting hundreds of miles into space. Lakes of liquid sulfur bubble across its landscape. The reason for all this activity is Io's closeness to Jupiter. The pull of the planet's gravity raises tides inside the moon, which produce heat and cause eruptions.

geyser

sulfur lake

surface of Io

JUPITER DATAFILE

Distance from Sun	484 million miles
Diameter	88,860 miles (11.2 Earths)
Mass	318 Earths
Day (time taken to spin once)	9 Earth hours 50 minutes
Year (time taken to orbit the Sun)	11 Earth years 314 days
Surface temperature	–238°F
Moons	16
Maximum magnitude	(*see page 45*) –2.7

Jupiter

Earth

◀ **Jupiter has a thin ring** around it, like Saturn's but much smaller. Jupiter's ring is very faint—even closeup it can hardly be seen.

▶ **The Great Red Spot** is like a hurricane in Jupiter's atmosphere, sometimes pulling in smaller storms, which show as white spots.

Patterns on Jupiter

Across the planet are dark bands, which change in darkness, size, and position from time to time. These are separated by lighter bands, called zones. A giant raft of colored gas, called the Great Red Spot, lies in the planet's southern hemisphere. Three times the size of Earth, it looks very different from year to year. Sometimes we can hardly see it against the creamy color of the rest of Jupiter, and sometimes it is almost brick red.

Like the wind patterns in Earth's atmosphere, the bands and zones on Jupiter are caused by the Sun's heat. This creates regions of rising and falling air, similar to the circulation of water boiling in a pan. Jupiter's rapid spin pulls these regions parallel with the equator, with the wind blowing one way along a band and the other way along the neighboring zone. Moving in opposite directions, the winds swirl around where they meet. When the Galileo probe descended through Jupiter's atmosphere, it was battered by 400 mph winds.

In 1994, the remains of a comet plunged into the planet. When the largest pieces hit Jupiter, they left black stains that lasted for a few months on the planet's cloud layers.

▲ **The Galileo spacecraft** reached Jupiter in December 1995. A probe from the main craft entered the planet's atmosphere, sending back information as it fell. Then the main craft went into orbit around Jupiter and sent back pictures of the planet and its satellites.

Saturn

Saturn Datafile

Distance from Sun
887 million miles

Diameter
74,900 miles (9.4 Earths)

Mass
95 Earths

Day (time taken to spin once)
10 Earth hours 14 minutes

Year (time taken to orbit the Sun)
29 Earth years 167 days

Surface temperature
–290°F

Moons
18

Maximum magnitude
(see *page* 45) –0.2

Earth

Saturn

The rings around Saturn make it everybody's favorite planet to look at through a telescope.

Saturn is second only to Jupiter in size and, like Jupiter, is made mostly of hydrogen and helium. It moves slowly, taking nearly 30 years to orbit the Sun. Like Jupiter, however, it spins rapidly—in 10 hours and 14 minutes—and it is even more flattened at the poles. Although Saturn is not much smaller than Jupiter, it has only about one third of the mass. This makes Saturn lighter than water—if you could drop it into a giant ocean, it would float!

Unlike Jupiter's constantly changing cloud belts, Saturn is relatively calm. A storm does break out from time to time, usually when Saturn is closest to the Sun in its orbit. From Earth, we see the storm only as a white spot on the planet.

▲ **The Hubble Space Telescope** took this picture of Saturn. Even the wide, dark gap in the rings, called the Cassini Division, has rings within it.

Rings and moons

At first glance there seem to be three rings forming a disk around Saturn. In fact, there are lots of rings. Each one is made up of many individual rings, which make it look grooved.

When the outer edge of the disk of rings is facing toward Earth, we can hardly see the rings at all—although the brightest of them are 170,000 miles wide, they can only be a mile or so thick. They are made up of millions of tiny fragments, mostly chunks of ice about the size of a snowball with perhaps a piece of rock in the center. The largest chunks are probably the size of a house while the smallest are as tiny as specks of dust. Each piece of material in a ring is in its own separate orbit around Saturn.

As well as the three broad rings, there are other fainter or narrower ones. Right at the outer edge of the ring system lies the strangest ring of all, called the F-ring. It has three or four separate ringlets, which are twisted or braided like pigtails along part of their length.

This effect is caused by two small moons, named Prometheus and Pandora, which orbit Saturn on

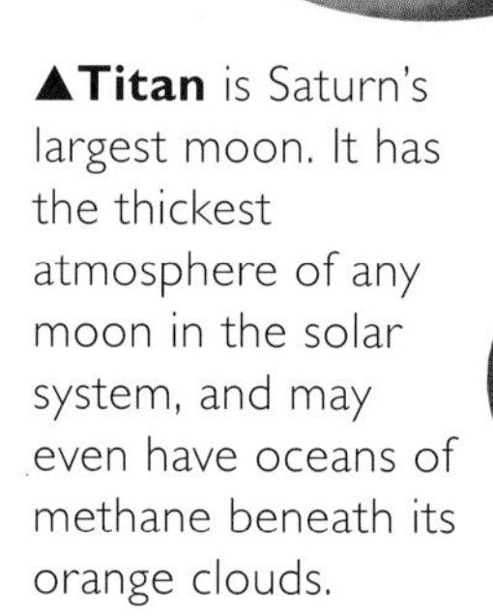

▲**Titan** is Saturn's largest moon. It has the thickest atmosphere of any moon in the solar system, and may even have oceans of methane beneath its orange clouds.

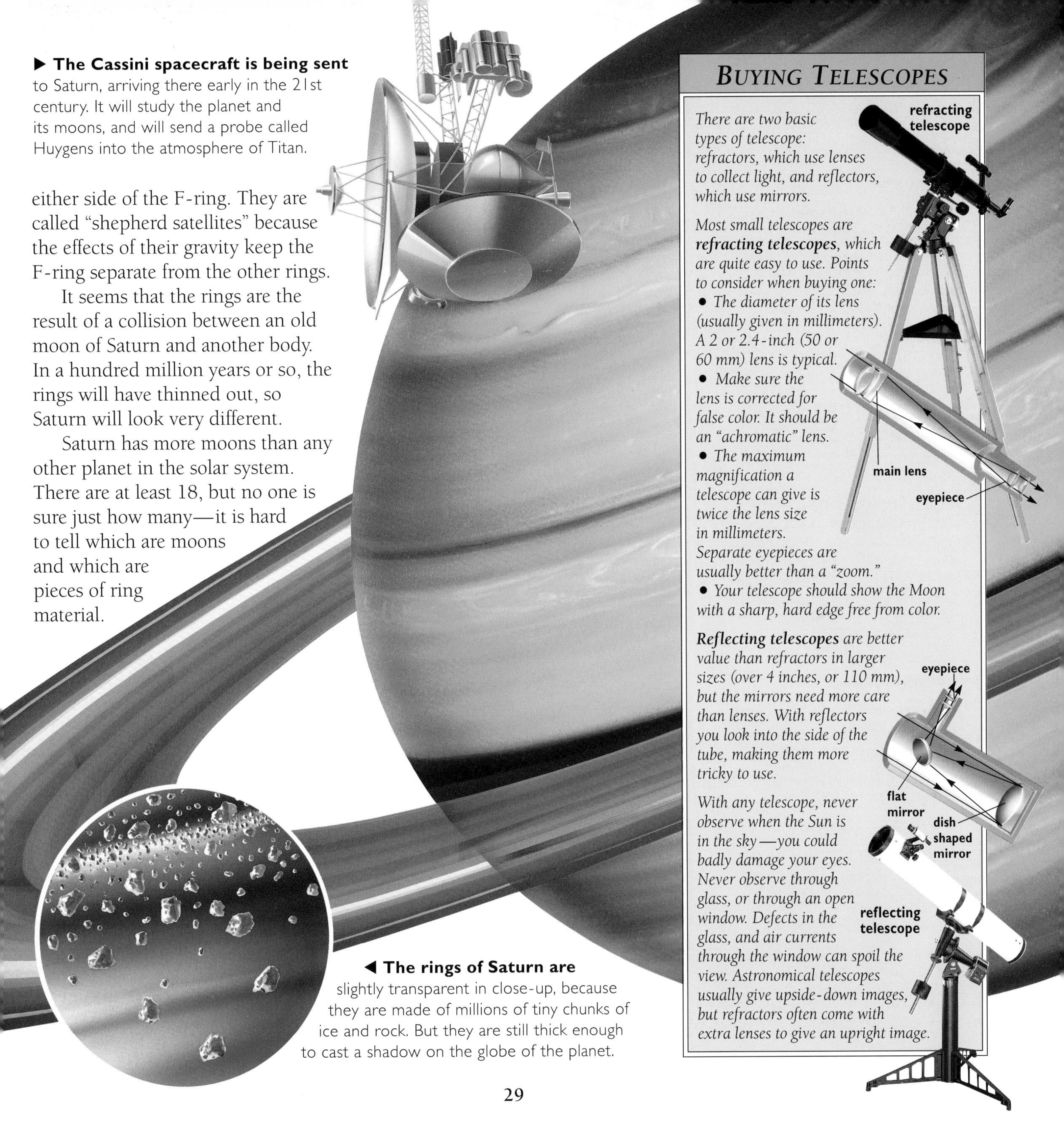

▶ **The Cassini spacecraft is being sent** to Saturn, arriving there early in the 21st century. It will study the planet and its moons, and will send a probe called Huygens into the atmosphere of Titan.

either side of the F-ring. They are called "shepherd satellites" because the effects of their gravity keep the F-ring separate from the other rings.

It seems that the rings are the result of a collision between an old moon of Saturn and another body. In a hundred million years or so, the rings will have thinned out, so Saturn will look very different.

Saturn has more moons than any other planet in the solar system. There are at least 18, but no one is sure just how many—it is hard to tell which are moons and which are pieces of ring material.

◀ **The rings of Saturn are** slightly transparent in close-up, because they are made of millions of tiny chunks of ice and rock. But they are still thick enough to cast a shadow on the globe of the planet.

BUYING TELESCOPES

There are two basic types of telescope: refractors, which use lenses to collect light, and reflectors, which use mirrors.

Most small telescopes are ***refracting telescopes****, which are quite easy to use. Points to consider when buying one:*

- *The diameter of its lens (usually given in millimeters). A 2 or 2.4-inch (50 or 60 mm) lens is typical.*
- *Make sure the lens is corrected for false color. It should be an "achromatic" lens.*
- *The maximum magnification a telescope can give is twice the lens size in millimeters. Separate eyepieces are usually better than a "zoom."*
- *Your telescope should show the Moon with a sharp, hard edge free from color.*

Reflecting telescopes *are better value than refractors in larger sizes (over 4 inches, or 110 mm), but the mirrors need more care than lenses. With reflectors you look into the side of the tube, making them more tricky to use.*

With any telescope, never observe when the Sun is in the sky—you could badly damage your eyes. Never observe through glass, or through an open window. Defects in the glass, and air currents through the window can spoil the view. Astronomical telescopes usually give upside-down images, but refractors often come with extra lenses to give an upright image.

Uranus and Neptune

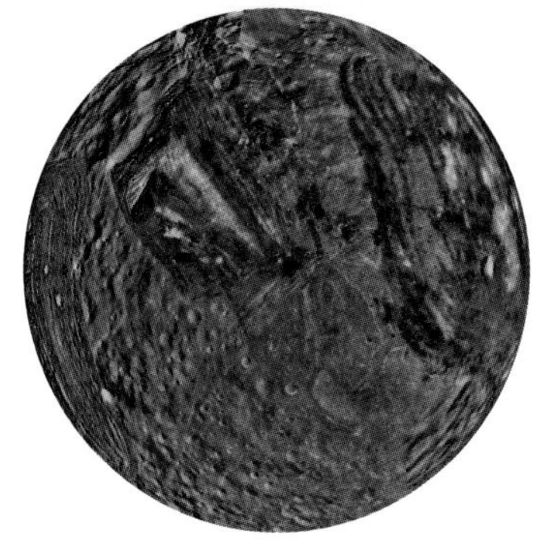

▲ **Miranda is the smallest** of Uranus's moons. It seems to have broken up in the past and pieced itself together again, leaving strange patterns on its surface.

Beyond Saturn lie the planets Uranus and Neptune. We knew little about them before they were visited by Voyager spacecraft in 1986 and 1989. In some ways they are like Jupiter and Saturn—they are much bigger than the rocky planets close to the Sun and are made of lighter materials.

Both planets are faint in our sky—Uranus is sometimes just bright enough to be seen with the naked eye, but you would need a telescope or binoculars to see Neptune. They move slowly around the Sun, so it takes some time to see any change in their positions.

Uranus and Neptune Datafile

Distance from Sun
Uranus: 1.784 billion miles
Neptune: 2.794 billion miles

Diameter
Uranus: 31,570 miles (4 Earths)
Neptune: 30,205 miles (3.8 Earths)

Mass
Uranus: 15 Earths
Neptune: 17 Earths

Day (time taken to spin once)
Uranus: 17 Earth hours 14 minutes
Neptune: 19 Earth hours 12 minutes

Year (time taken to orbit the Sun)
Uranus: 84 Earth years 4 days
Neptune: 165 Earth years

Surface temperature
Uranus: –346°F
Neptune: –360°F

Moons
Uranus: 15
Neptune: 8

Maximum magnitude (*see page 45*)
Uranus: 5.5
Neptune: 7.8

Uranus

For centuries, people knew of only five planets—Mercury, Venus, Mars, Jupiter, and Saturn. Then, in 1781, a musician and amateur astronomer named William Herschel was looking at the stars with a home-made telescope when he found a new planet: Uranus.

Like Jupiter and Saturn, Uranus has no solid surface. Its outer layers are made of hydrogen, helium, and methane. The methane gives Uranus a slight blue-green color, but the planet has few markings—just faint bands. As you go deeper, there is more and more water. At first, there is a thin mist, which gets thicker until it becomes a warm, dark ocean. Near the center of the planet, this ocean gets increasingly hot and sludgy until it becomes solid rock.

▶ **Uranus has thin rings,** like Saturn's but much narrower and fainter. The ice particles in them are the color of coal, while Saturn's are like snow. Uranus's rings are probably much older than Saturn's.

Unlike any other planet, Uranus spins on its side—the other planets have poles that point more or less upward and downward, while Uranus's poles point sideways.

Uranus has 15 moons. The smallest, Miranda, has ice cliffs 12 miles high, which is enormous for a world only 300 miles across.

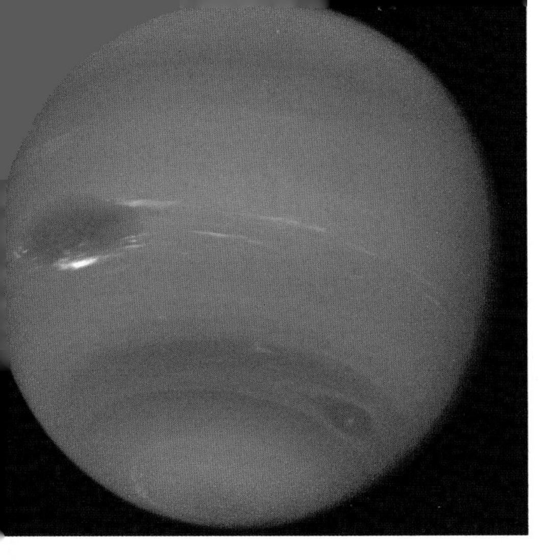

◀ **When Voyager 2 took** this picture of Neptune in 1989, the planet had a Great Dark Spot (*left of center*) in its southern hemisphere. This had gone by 1994 when the Hubble Space Telescope sent back pictures of the planet.

▶ **Each Voyager** spacecraft used the pull of gravity of giant Jupiter to increase its speed and send it on its way to Saturn and beyond.

Saturn
Jupiter
Earth
Voyager 1
Uranus
Voyager 2
Neptune

Voyages into the Unknown

The most successful mission to the planets was Voyager. Two spacecraft sent back close-up pictures of Jupiter, Saturn, Uranus, Neptune, and many of their moons.

Voyagers 1 and 2 were launched within a few days of each other in 1977. Voyager 1 sent back pictures of giant Jupiter as it flew past the planet in 1979, and then went on to visit Saturn in 1980.

Voyager 2 took a slower route to Jupiter, reaching it four months after Voyager 1, and then arriving at Saturn in 1981. It passed Uranus in 1986, and finally, after 12 years in space, it reached Neptune in 1989. Both Voyagers are now heading out into space beyond the planets.

Each spacecraft carries cameras as well as detectors to measure such things as magnetic fields and solar wind particles (see page 19) in space.

Because the Voyagers may some time in the future be found by other civilizations, they carry phonograph players and records with different sounds from Earth, together with pictures of its peoples and cultures.

Neptune

Uranus did not seem to move in its orbit around the Sun in the way astronomers expected it to. This suggested that there was another planet even farther out. French and British mathematicians worked out where this new planet should be found, and in 1846 Neptune was spotted by a German astronomer.

Neptune shows more markings than Uranus, although it, too, is mainly made of water with an atmosphere of hydrogen, helium, and methane. Dark spots come and go, while bright clouds of methane hang high up in its atmosphere.

Winds rage around Neptune at over 600 mph—much faster than any hurricane on Earth.

Neptune has some strange moons. The largest, Triton, is the only moon of any planet that orbits in the opposite direction to the direction in which the planet spins. Although Triton is one of the coldest places in the solar system at –455°F, it has geysers. Scientists think that its surface of clear ice acts like the glass of a greenhouse and heats gas just below the moon's surface.

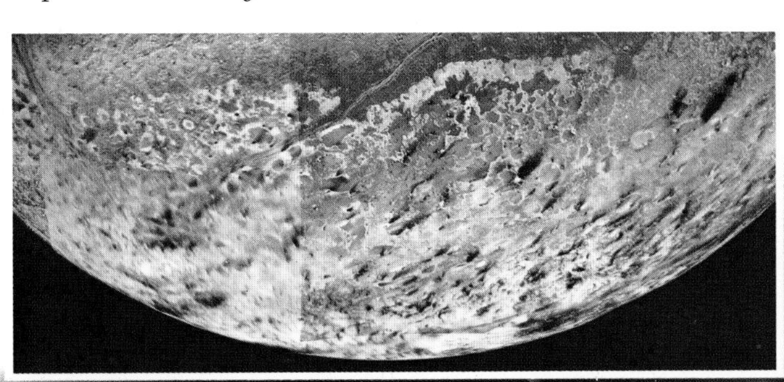

▶ **Triton's surface** is covered with clear, pinkish ice. Frozen nitrogen below the ice is heated, possibly by the distant Sun, and gushes from cracks forming five-mile-high geysers.

Pluto and beyond

At the very edge of the solar system lies the tiny planet Pluto, as well as millions of smaller objects that make up what is called the Kuiper Belt.

Little of the Sun's heat reaches this part of the solar system, so it is mostly ice. When the planets formed (*see page 11*), the heavier materials stayed close to the center of the solar system to form the rocky planets and the asteroids. Farther out, the main materials were ice and dust.

Some of this ice and dust formed the large planets Uranus and Neptune, while the rest made smaller chunks. Pluto is a strange mixture. Its orbit is at a different angle to the orbits of the other planets, and it is smaller than several of the solar system's moons. Some astronomers believe that it is not a true planet, but is just the largest known member of the Kuiper Belt.

orbit of Pluto
Charon
center of gravity
Pluto
orbit of Charon

▲ Pluto's moon, Charon, is half the size of Pluto. The two make a double planet, with both orbiting a point between them, known as the center of gravity. Their orbits are like the ends of a twirling baton that has one end heavier than the other.

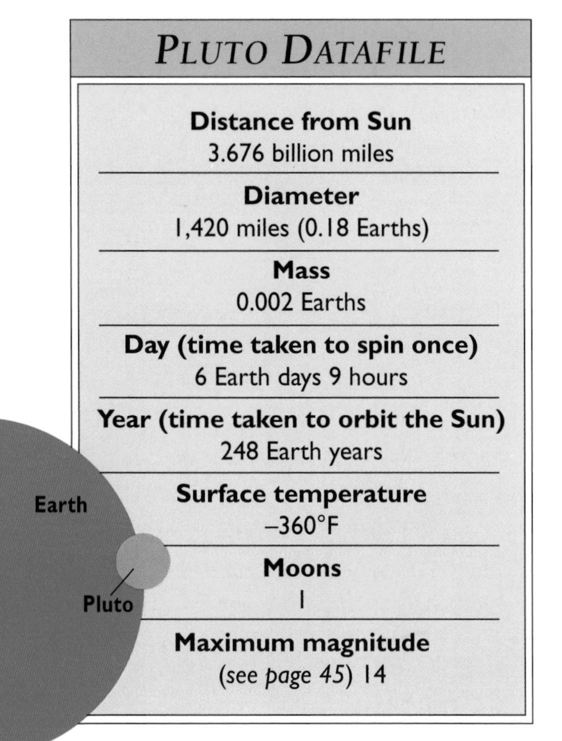

Pluto Datafile

Distance from Sun
3.676 billion miles

Diameter
1,420 miles (0.18 Earths)

Mass
0.002 Earths

Day (time taken to spin once)
6 Earth days 9 hours

Year (time taken to orbit the Sun)
248 Earth years

Surface temperature
–360°F

Moons
1

Maximum magnitude
(see *page* 45) 14

Pluto

In the same way that Neptune was found because of its effect on the movement of Uranus, Pluto was discovered because Neptune moved from the orbit astronomers expected it to have. The mystery planet was found by Clyde Tombaugh, who looked at photographs of millions of stars until, in 1930, he found one that moved more like a planet than a star. This was Pluto.

Pluto has one moon, called Charon, which orbits the planet in 6 days 9 hours. This is the same time that it takes Pluto to spin once, so Charon always stays above the same point on Pluto's surface.

Pluto comes closer to the Sun than Neptune for 20 years of its 248-year orbit. The planet changes as it moves from its closest point to the Sun (within about three billion miles) to its farthest point (about five billion miles). When it is at its closest, Pluto has a thin atmosphere of methane, nitrogen, and carbon dioxide. As it moves away, these gases freeze onto its surface. At its farthest from the Sun, Pluto is completely frozen.

Beyond Pluto

In 1992, another object was found orbiting the Sun at about four billion miles away. Called 1992 QB1, it appears to be only about two hundred miles across. More of these little objects are still being found.

Astronomers call them Trans Neptunian Objects, or TNOs ("trans" means "on the other side of"), and they may be the closest members of a belt of material at the outer edge of the solar system.

The discovery of these small, icy objects supports astronomer Gerard Kuiper's idea that many comets must orbit just beyond Pluto. This area is now known as the Kuiper Belt. The body of a comet is simply an icy chunk a few miles across—it looks like a comet only when it comes close to the Sun (*see pages 20–21*). So there may be little difference between TNOs, comets, and the moons of the outer planets—they just have different orbits.

The Kuiper Belt may contain millions of icy bodies, each too small to be called a planet. Some people think that Pluto and Charon are members of the Kuiper Belt, and call them "super-comets" or "ice dwarfs."

▲ **Beyond the Kuiper Belt lies the Oort Cloud.** This contains more icy fragments, which orbit the Sun in every direction. Some brilliant comets come from the Oort Cloud.

▶ **The Kuiper Belt** surrounds the icy outer planets, just as the asteroid belt surrounds the inner rocky planets. Both belts are probably made up of material that did not form into full-sized planets when the solar system was born.

▼ **Seen from Pluto, Charon** is always in the same place above the planet's surface. The Sun looks like a star in Pluto's sky, but it is 250 times brighter than the Earth's Full Moon.

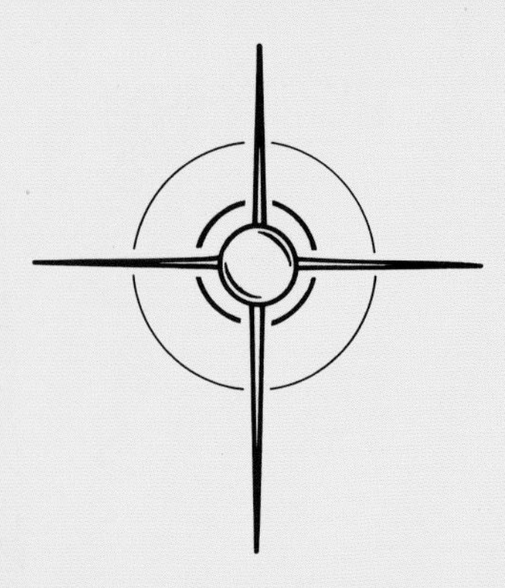

▲ **Stars have different colors,** which show up more in photographs than when seen with the naked eye or through binoculars.

THE STARS

APART FROM THE SUN, ALL STARS ARE SO FAR AWAY THAT, IN MOST cases, we see them only as tiny points of light in the night sky. But astronomers have learned a lot about what stars are and how they behave simply by looking at their light.

What star colors tell us

The color of a star is usually a guide to its temperature and true brightness (this is the brightness with which it shines in space, not the brightness it appears to have when we see it from Earth). The coolest and dimmest stars are reddish. Warmer, brighter stars are yellow or white, and the hottest and brightest ones are blue-white. Think of the color of metal as it is heated—first it glows dark red; then, as it gets hotter, it gets brighter and changes color to orange, yellow, and eventually brilliant white.

The colors of stars can tell us their masses, too. Blue stars are often huge, with masses up to 60 times bigger than the Sun's. Red stars are usually small, with only around a tenth of the Sun's mass.

There are exceptions. Some red stars, called red giants, are very bright, and some white stars, called white dwarfs, are faint. Both red giants and white dwarfs are particular stages in a star's life (*see pages 40–41*).

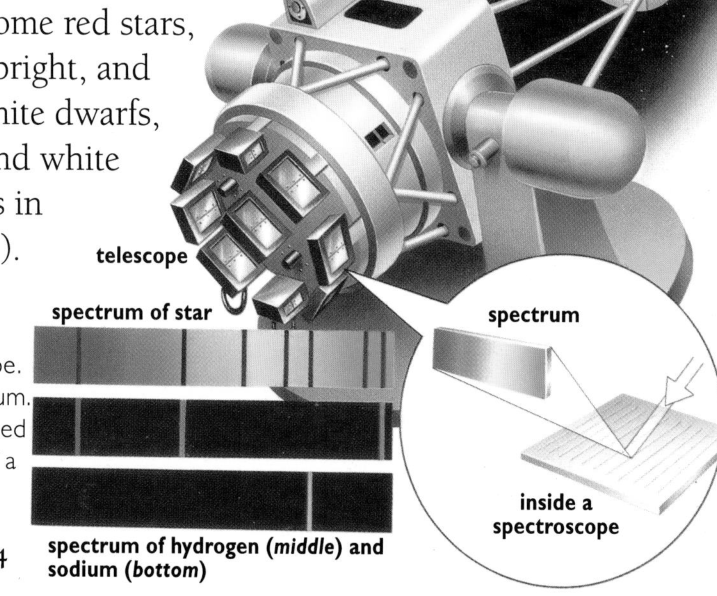

▶ **Light can be studied** using a spectroscope attached to a telescope. Inside, the light is split into a spectrum. A star's spectrum is a rainbow crossed by dark lines, while the spectrum of a gas is dark, with bright lines.

Types of stars

Astronomers group stars into types according to their color, and label each type with a letter: O, B, A, F, G, K, or M. The hottest blue stars are types O and B. Types A, F, and G are white and less hot. Type K is yellow. Redder stars are type M, and are the coolest. These groups are broken down further using numbers from 0 (the hottest) to 9 (the coolest). The Sun, for example, is a white G2 star, though the Earth's atmosphere sometimes makes it look yellow.

O-type star
radius: 12 Suns
true brightness: 500,000 Suns
surface temperature: 53,500°F

B-type star
radius: 6 Suns
true brightness: 800 Suns
surface temperature: 19,000–53,500°F

A-type star
radius: 2 Suns
true brightness: 50 Suns
surface temperature: 13,000–19,000°F

F-type star
radius: 1.3 Suns
true brightness: 6 Suns
surface temperature: 10,300–13,000°F

G-type star (like the Sun)
radius: 1 Sun
true brightness: 1 Sun
surface temperature: 8,500–10,300°F

K-type star
radius: 0.74 Suns
true brightness: 0.16 Suns
surface temperature: 5,800–8,500°F

M-type star (red dwarf)
radius: 0.3 Suns
true brightness: 0.01 Suns
surface temperature: 5,800°F

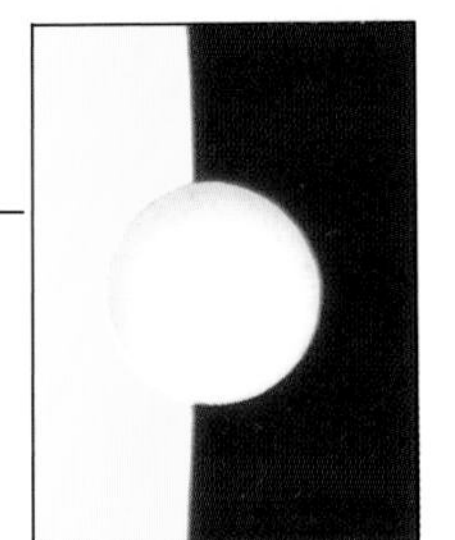

the size of a white dwarf compared to the Sun

A red giant star may be hundreds of times larger than an ordinary star (*see pages 40–41*). Betelgeuse, for example, is nearly 1,000 times the size of the Sun.

Clues in starlight

Starlight carries even more clues to what stars are. White light is made up of all the colors of the rainbow mixed together. It can be split into its colors using a solid triangle of glass, called a prism. As the light passes through the prism, the colors separate because each one is bent by the glass at a different angle. Instruments that split starlight use thousands of very fine grooves cut into a sheet of glass or metal.

When light is split into its colors, it makes a spectrum. This is a band with blue at one end and red at the other, with all the other colors in between. A spectrum of starlight has thin, dark lines along it, as if some of the colors are missing. This is exactly what happens—cool gases around each star absorb certain colors so that they do not reach Earth and do not appear in the spectrum. If a gas is hot, it shines only with particular colors, and its spectrum is dark with these colors showing as bright bands. Hot sodium gas, like that used in some streetlights, glows yellow. The same yellow appears in its spectrum. If a star's spectrum shows a dark line (called a spectral line) where that yellow should be, we know that the star contains sodium.

By comparing the positions of spectral lines over a period of time, astronomers can find out whether a star is moving toward or away from us, because the lines shift slightly (*see box*). The darkness and width of spectral lines also give clues to a star's true brightness, which can show whether it is a giant star or a dwarf. This method of studying light is called spectroscopy.

DOPPLER SHIFT

A spectrum can show how fast a star is moving, and whether it is moving toward or away from us. If the star is moving away from us, its spectral lines will move slightly toward the red end of the spectrum. The higher the speed, the farther these lines will move. If the star is moving toward us, the shift is toward the blue end of the spectrum. This change is called the Doppler shift.

▼ **The positions of the spectral lines** of a star are compared with those of a gas on Earth. **A** shows the star is moving away from us. **B** shows it is moving toward us.

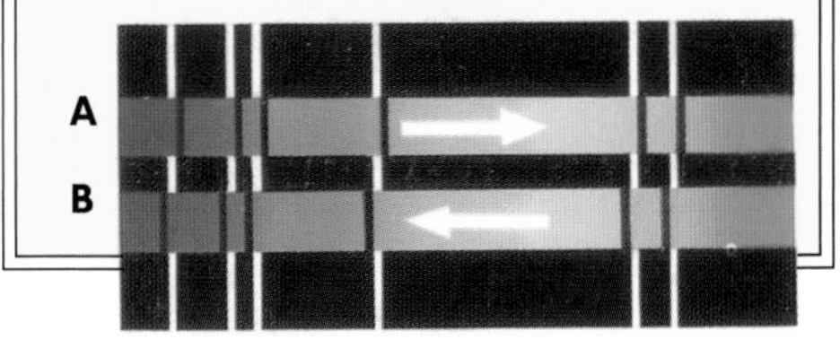

Measuring star distances

You might think that the brighter stars in the sky are near, and the fainter ones are farther away. But brightness is a poor guide to distance. The nearest star to the Sun, Proxima Centauri, is so faint that we cannot see it with the naked eye. Yet some of the brightest stars we can see are very distant.

The only accurate way of measuring a star's distance is by using what is called parallax. This is the movement an object seems to make in its position when it is seen from two different viewpoints. Our eyes use parallax all the time to tell us distances. Try looking at a nearby object with one eye closed, then with the other eye closed. The position of the object against its background changes slightly. Our brain uses this change to decide how far away an object is.

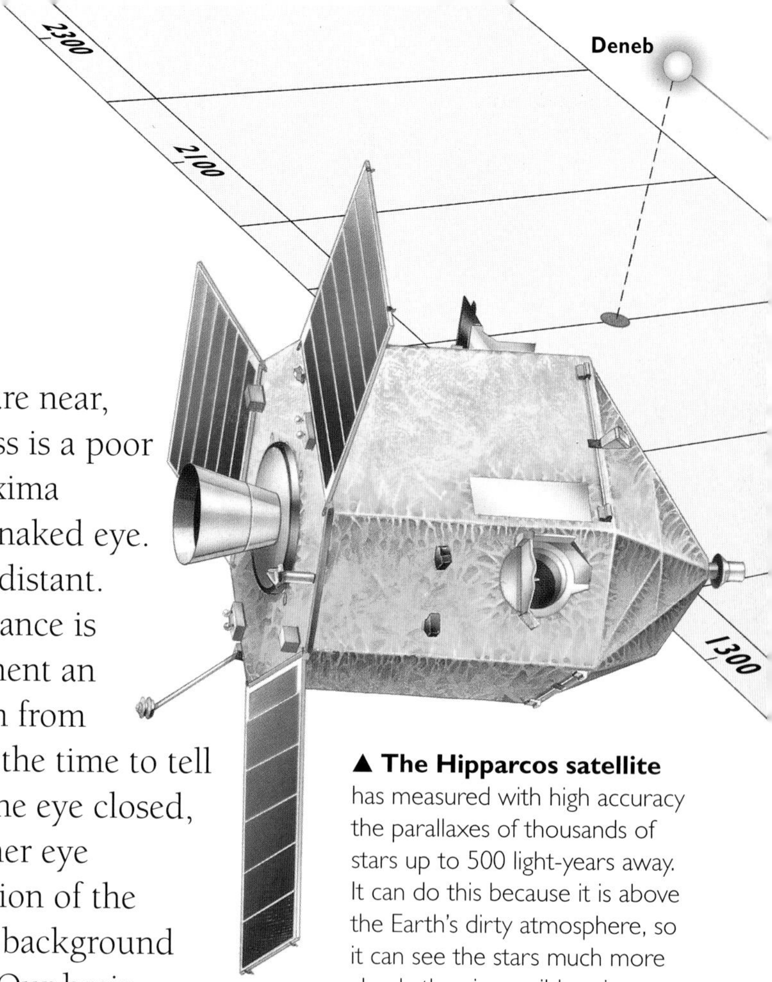

▲ **The Hipparcos satellite** has measured with high accuracy the parallaxes of thousands of stars up to 500 light-years away. It can do this because it is above the Earth's dirty atmosphere, so it can see the stars much more clearly than is possible using telescopes on the Earth's surface.

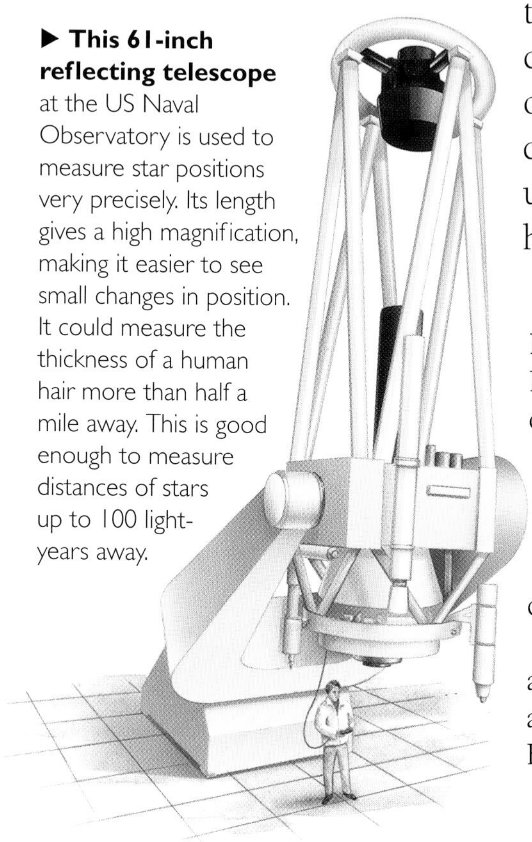

▶ **This 61-inch reflecting telescope** at the US Naval Observatory is used to measure star positions very precisely. Its length gives a high magnification, making it easier to see small changes in position. It could measure the thickness of a human hair more than half a mile away. This is good enough to measure distances of stars up to 100 light-years away.

Distant objects

If you try to use parallax on an object more than 100 yards away, you will not see any movement. Your eyes would have to be much wider apart. Only if you moved from side to side would you see a change in the object's position.

To find out the distances of stars, astronomers need viewpoints as far apart as possible. The orbit of the Earth around the Sun gives them the greatest distance between viewpoints. By observing a nearby star from one side of Earth's orbit and then, about six months later, from the other side, astronomers can measure a slight change in the star's position. From this, they learn its distance from Earth.

Parallax works only for stars less than a few hundred light-years away. For a more distant one, astronomers first decide what type of star it is,

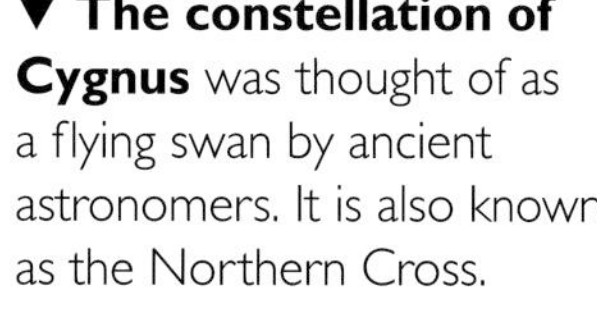

▼ **The constellation of Cygnus** was thought of as a flying swan by ancient astronomers. It is also known as the Northern Cross.

USING PARALLAX

To measure the parallax of a star, astronomers take photographs of it against the background of more distant stars. As the Earth moves around the Sun, the star's position against the background will appear to change slightly. The closer the star is to Earth, the greater this change will be. Astronomers take many photographs over months or even years to measure the star's distance accurately.

▶ **Seen from either side of Earth's orbit,** stars **A** and **B** each move slightly against the star background. The closer star, **B**, moves more. The actual movement is so small that it shows up only after careful measurements of the stars' positions.

such as type B or type F (*see pages 34–35*). This tells them how brightly the star shines in space. Then they can work out how far away it must be to look as bright (or as faint) as it does from Earth. For example, if a star is type B, astronomers compare its brightness with another, closer, type B star whose distance is known. The fainter the distant star appears, the farther away it must be.

▼ **The bright stars of Cygnus** are all at different distances from Earth. Its brightest star, Deneb, is the most distant in the group. The star 61 Cygni is the closest, but is quite faint. It was the first star to have its distance measured.

▲ **The true distances** of the bright stars of Cygnus are shown in this diagram. It is impossible to judge their distances just by looking at them in the night sky.

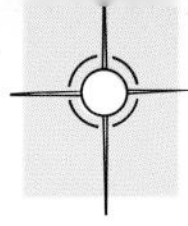

Life of a star

Stars are born, change as they get older, and die. They form out of clouds of gas and dust in space. As stars are born, planets may form around them.

The birth of a star begins with wispy gas. This gas is mostly hydrogen, with some helium and other elements and dust mixed in. To begin with, there is no obvious center to the clouds of gas, but after millions of years, perhaps by chance, some of the particles may begin to clump together.

When a clump forms, its gravity starts to pull in gas and dust from the surrounding cloud. As the clump grows in mass, the pull of its gravity increases and attracts more material. The particles of gas and dust are pulled inward toward the center of the clump where they become more and more tightly packed, so that the center also gets smaller.

Double Stars

Many stars are not single, but are made up of two stars which formed together and which orbit each other. The spectral lines of a double star split in two when the two stars move in opposite directions. The Doppler shift (see page 35) shows each star's speed and size of orbit, and because heavy stars orbit differently to light stars, astronomers can work out each star's mass.

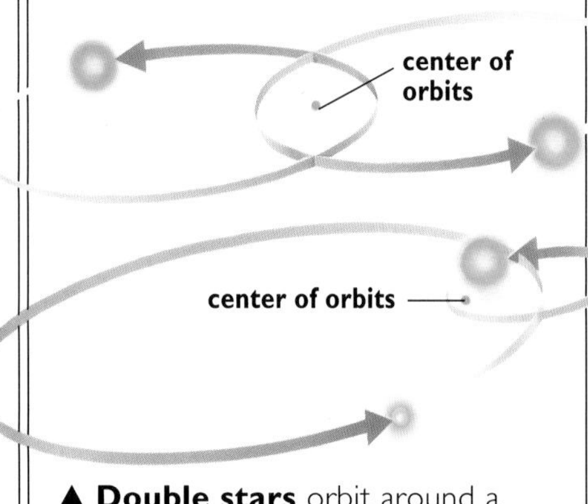

▲ **Double stars** orbit around a point between them. If the two stars have the same mass, they will have the same size of orbit (*top*). If one is smaller, its orbit is larger (*above*).

▼ **The Hubble Space Telescope** (HST) can take extremely sharp pictures of stars. It was launched in 1990 aboard a space shuttle, and its instruments can be changed while it is in space.

A star is born

As the particles are crammed together in the center of the clump, they jostle against one another. This makes them heat up. They get hotter and hotter until the center starts to glow. It is now known as a protostar. The most important event in the protostar's life is when the temperature at its center reaches about 27 million°F. This is hot enough for nuclear energy to be made, just as happens inside the Sun (*see page 13*).

▶ **Stars form inside huge clouds** of gas and dust like this one, called the Eagle Nebula. You can see some new stars at the ends of the fingers of gas.

At first, the protostar spins slowly inside its cloud. But, as it shrinks in size, it speeds up, just as an ice skater spins faster by pulling in his or her arms. This flattens the surrounding cloud into a thick disk. Separate clumps inside this disk may be big enough to form other stars. If these clumps are small, they might form planets.

The protostar is still surrounded by the gas and dust from which it was born, and it cannot be seen from outside. But as it starts to shine, its heat blows the remaining gas and dust away, leaving behind only what has already formed into other stars or planets. The heat from nearby stars can also help to do this. Finally, after millions of years, all the material around the star has gone, and it settles down to shine steadily.

Mass and brightness

The more massive the star, the stronger its gravity will press inward on its center, or core. The star will use up its store of hydrogen more quickly and it will shine more brightly, but its life will be shorter.

Big O-type blue stars use up their hydrogen fuel so fiercely and shine so brightly that they will start to run out of fuel within only five million years. A G-type star like the Sun, however, shines steadily for about 10 billion years. Faint M-type red dwarfs will last several times as long as this.

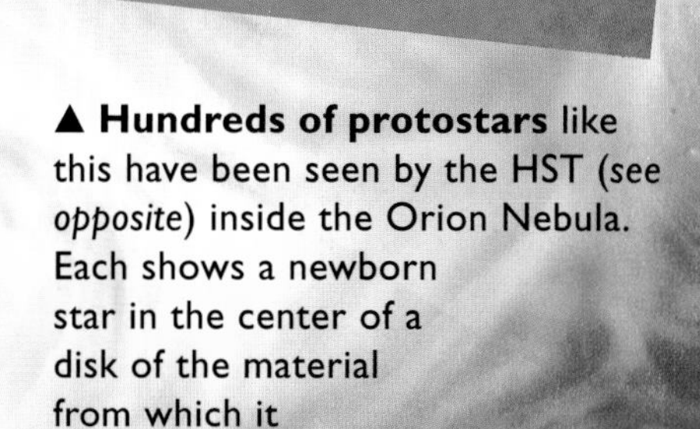

▲ **Hundreds of protostars** like this have been seen by the HST (*see opposite*) inside the Orion Nebula. Each shows a newborn star in the center of a disk of the material from which it was born.

▶ **Star formation** can be triggered in a cloud of gas and dust by the shock waves from a supernova explosion (*see pages 40–41*), which can make particles clump together. After millions of years, (*from left*) a cluster of new stars forms hidden inside the cloud. Eventually, the energy from these and nearby stars blows away the surrounding material and the cluster can be seen.

▶ **A star shines** once it has blown away its cloud of gas and dust. The heat from nearby stars helps to do this. Material in the disk around the new star may form planets, before the rest is blown away.

1 Nearby stars help to blow away gas and dust.

2 The new star also blows away material.

3 Planets and other stars may begin to form from the gas and dust.

4 A system of planets has formed around the star.

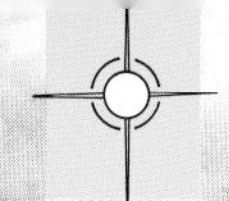

Dying stars

A star has only a limited amount of hydrogen fuel. When its supply begins to run out, the star's life is coming to a close. Some stars die in a spectacular explosion while others simply fade away. The final stages in a star's life depend on its mass.

In all stars, there is a delicate balance of forces. Gravity is always trying to pull the star in on itself, but as long as the star keeps producing energy in its center, turning hydrogen into helium, the outward flow of energy balances the inward pull of gravity. When the hydrogen runs low, gravity starts to win and the center of the star gets smaller again, just as it did when the star was born. The particles at the center jostle together and heat up even more—from 27million°F to a fierce 180 million°F.

▲ **The red giant star Betelgeuse,** photographed by the Hubble Space Telescope. The temperature of the star's surface is 9,000°F, while the brighter spot seen at its center is some 3,000°F hotter.

The final stages

Once the hydrogen in the center of the star is used up, the helium that is left becomes the fuel to produce another element, carbon. The star now swells to many times its original size, and because this spreads the flow of energy over a huge area, the surface gets cooler. The star is now a red giant.

▲ **The Egg Nebula** surrounds a star which was a red giant until a few hundred years ago. It is puffing off shells of gas.

Some of the carbon made in the red giant gets mixed with its outer layers, which are slowly blown away from the star as the core runs out of fuel. This makes red giants a major source of the clouds of dust in space. All the carbon on Earth, including that inside your body, was formed in space inside stars.

When the red giant runs out of fuel, the star's time is up. Its core can no longer produce energy, so it collapses. A star of average mass will have lost its outer layers in space, and the remains of the center become a tiny, dim white dwarf. For a while, the lost outer layers form a glowing shell around the star. This is called a planetary nebula because it makes the star look like a distant planet. After billions of years, the white dwarf fades away, becoming a black dwarf.

A star about five times the Sun's mass dies much more dramatically, in a supernova. For a few days the star can be as bright as billions of stars put together.

In this incredible blast, elements such as iron, silver, and gold are created and scattered through space. So, when you see something made of silver or gold, remember that the metal was created in a supernova.

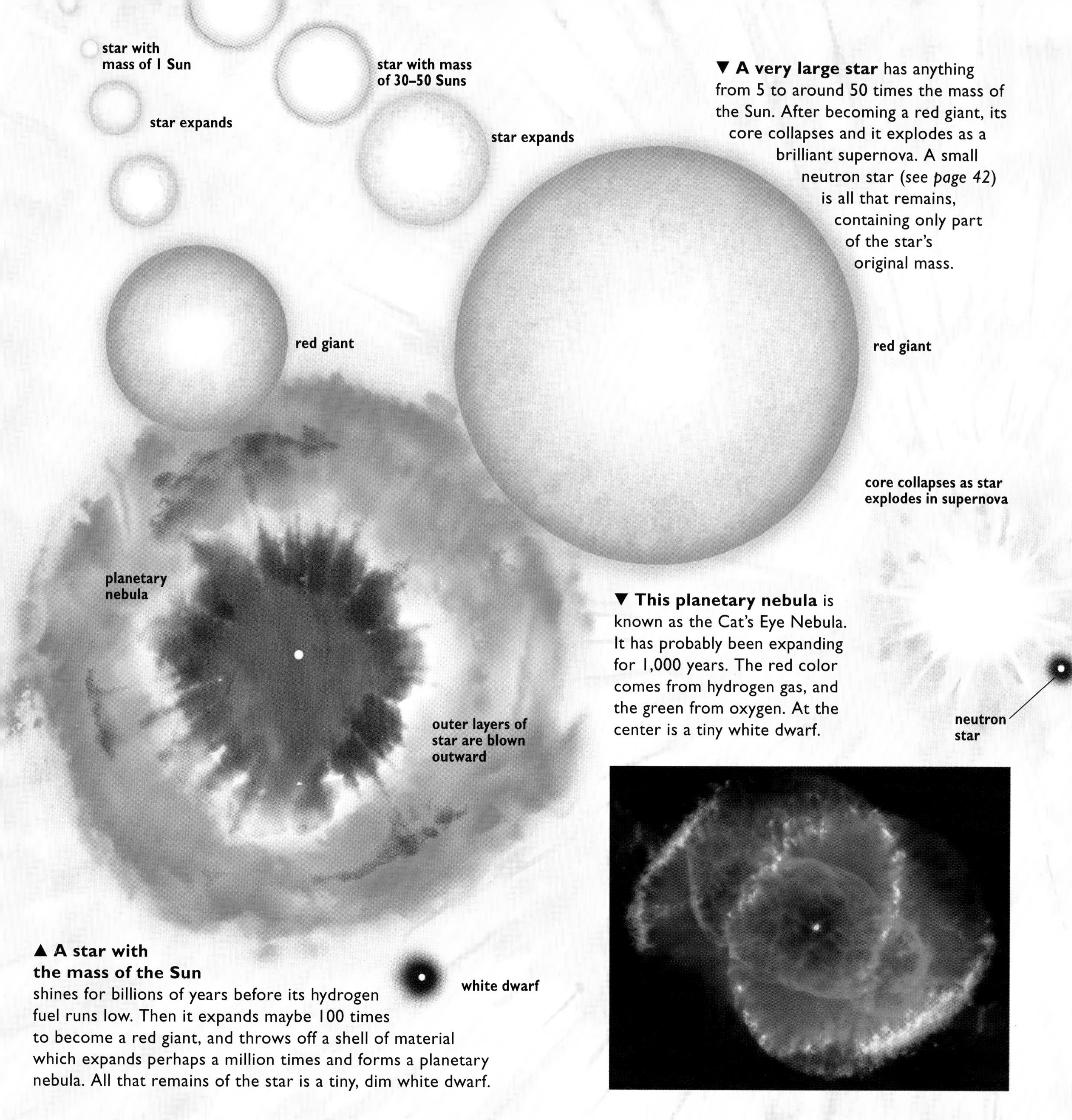

▼ **A very large star** has anything from 5 to around 50 times the mass of the Sun. After becoming a red giant, its core collapses and it explodes as a brilliant supernova. A small neutron star (*see page 42*) is all that remains, containing only part of the star's original mass.

▼ **This planetary nebula** is known as the Cat's Eye Nebula. It has probably been expanding for 1,000 years. The red color comes from hydrogen gas, and the green from oxygen. At the center is a tiny white dwarf.

▲ **A star with the mass of the Sun** shines for billions of years before its hydrogen fuel runs low. Then it expands maybe 100 times to become a red giant, and throws off a shell of material which expands perhaps a million times and forms a planetary nebula. All that remains of the star is a tiny, dim white dwarf.

Dead stars: pulsars and black holes

axis

neutron star

▲ **A pulsar** is a rapidly spinning neutron star that beams radio waves from its magnetic poles. Because it usually spins on a different axis from its magnetic poles, the beams sweep around like those from a lighthouse.

radio beam

radio telescope on Earth

chart showing pulse of radio beam

▲ **As the beams** sweep across a radio telescope on Earth, they produce a stream of pulses in its receiver. A radio telescope picks up radio waves using a huge curved metal dish and reflects them into an antenna at its center.

Even the force of a great supernova explosion may not destroy a star completely. Part of it can still be left as a mysterious object called a neutron star. This may become a stranger object still—a black hole.

◀ **To find black holes,** astronomers use satellites, such as the RXTE, to look for X-rays coming from the gas surrounding them (see *opposite*).

A neutron star forms because gravity pulls the center of the massive star in on itself. The particles inside the atoms are crammed closer and closer together. The atoms cannot stand the strain and they collapse, forming just neutrons in the core. A sugar cube of this material would weigh a billion tons if it were on Earth.

Because the material has collapsed, the star also shrinks. Instead of being many thousands of miles across, it is now only about 10 or 15 miles in size.

▲ **A neutron star** is no bigger than a large city, yet it contains about as much material as a star like the Sun.

Pulsars

As the star shrinks, it spins ever faster. Instead of turning once every few weeks like a full-sized star, it may turn many times a second!

The pull of the star's magnetic field also gets stronger. The magnetic field picks up electrons from the surface of the star and they flow away out of the star's magnetic poles creating beams of radio waves. Astronomers can find these using radio telescopes, which detect radio waves rather than light waves.

As the star turns, its radio waves sweep across space. When a beam from a neutron star shines into a radio telescope, it produces a pulse in the telescope's receiver. This happens again and again as the star spins, making a steady stream of pulses. So these objects are known as pulsars.

Black holes

If the dying star is very massive, something even stranger happens. Because the weight of its outer layers is so great, it just goes on shrinking until it is infinitely dense (tightly packed) and far smaller even than the pupil in your eye.

Light is affected by gravity. As the dying star shrinks, the light has to struggle harder against the star's gravity to leave it. Once the star shrinks past a certain size, no light can leave it—in fact, nothing at all can get out of it. This is a black hole.

People often think that black holes suck in everything around them, but this is not quite true. Only if you were to go very close to a black hole would you fall into it. Any gas close to a black hole will be pulled first into a disk around it, which shines brilliantly as the gas then plunges in. So the area surrounding a black hole is anything but black!

▼ **If a black hole** is close to another star, gas will stream off the star into a disk around the black hole. The gas gets very hot as it plunges in, causing it to give off X-rays as well as light.

Formation of a Black Hole

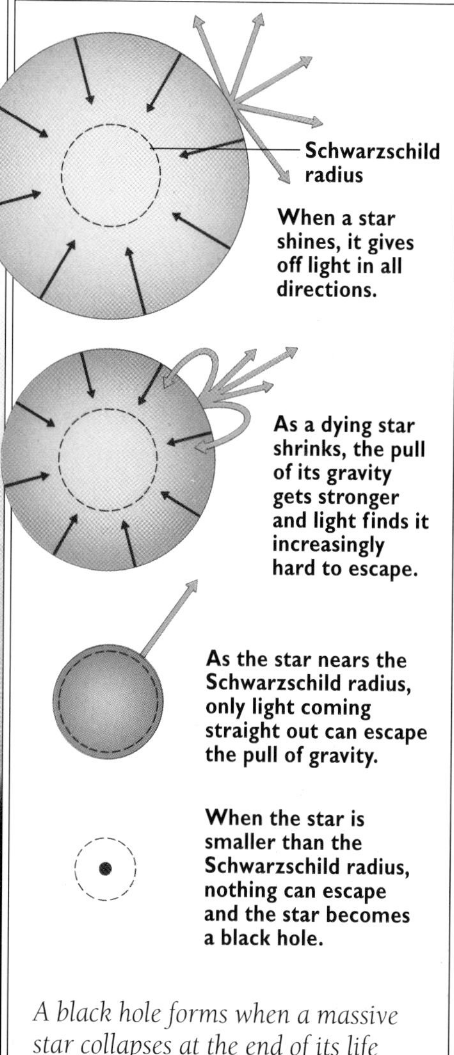

A black hole forms when a massive star collapses at the end of its life because it cannot produce any more energy (see pages 40–41). *Just as a rocket will fall back to Earth if it does not have enough power to beat the pull of gravity, light from a black hole will not be able to escape. The black hole itself is tiny, but the area from which light cannot escape—called the Schwarzschild radius—is usually several miles across.*

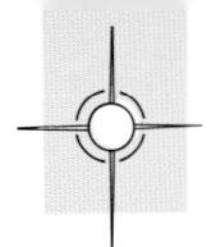

Star maps

Learning your way around the stars can seem difficult at first, but it does not take long to get the hang of it, and once you do, it is great fun. The secret is to learn the main star patterns first, then use these as landmarks to find others, much as you do when exploring any new area.

The main difficulty in learning the stars lies in the fact that different stars are seen through the year. As the Earth moves around the Sun, different parts of the heavens are visible. The Earth itself turns, so the stars appear to move slowly across the sky through the night.

▼ **We can think of the sky** as the inside of a huge ball around the Earth. The sky maps in this book have cut this ball into six pieces and opened it up (see *below*), making four main maps and two polar maps (north and south). On the maps, notice how the Sun's path around the sky, called the ecliptic (see *page 51*), becomes a curved line instead of a circle.

Using the star maps

The star maps on these pages make it easy to discover which stars you will see on a particular night. The sky has been divided up into six maps: there are four main maps, plus one each for the stars around the north and south poles of the sky.

The maps can be used anywhere in the world, but the instructions are given as if you are in the northern hemisphere. From the southern hemisphere, just turn the maps upside down. Then, for "south" read "north" and for "north" read "south." Near the equator, the stars on the main maps appear almost overhead, with half of both the northern and southern polar stars in the sky at the same time.

For any date, follow this step-by-step guide to the stars:

1 Choose the main map with the month in which you are observing. The map will show the stars in the sky looking southward at 9:45 p.m. standard time (10:45 p.m. if the clocks have been put forward for an hour of Daylight Saving Time). Wherever you are, you will be able to see the stars in the top half of each main map, but you may not be able to see the stars at the bottom.

2 To get an idea of scale, hold your open hand at arm's length. The distance between the tips of your thumb and your little finger gives you a 20-degree area of the sky. Compare this with the hand shown at right, which is drawn to the same scale as the maps.

3 The top of each main map overlaps the map of the northern polar stars (the overlap point is high in the sky). Turn the northern star map so that the same month is at the top. Then the map will be the right way around to show you the stars in the sky as you look north.

4 On each main map there are two or more main star groups with gray lines drawn on to help you spot the pattern. Find these patterns first, then go from star to star to find the others. This is called "star-hopping."

NORTH POLAR STAR MAP

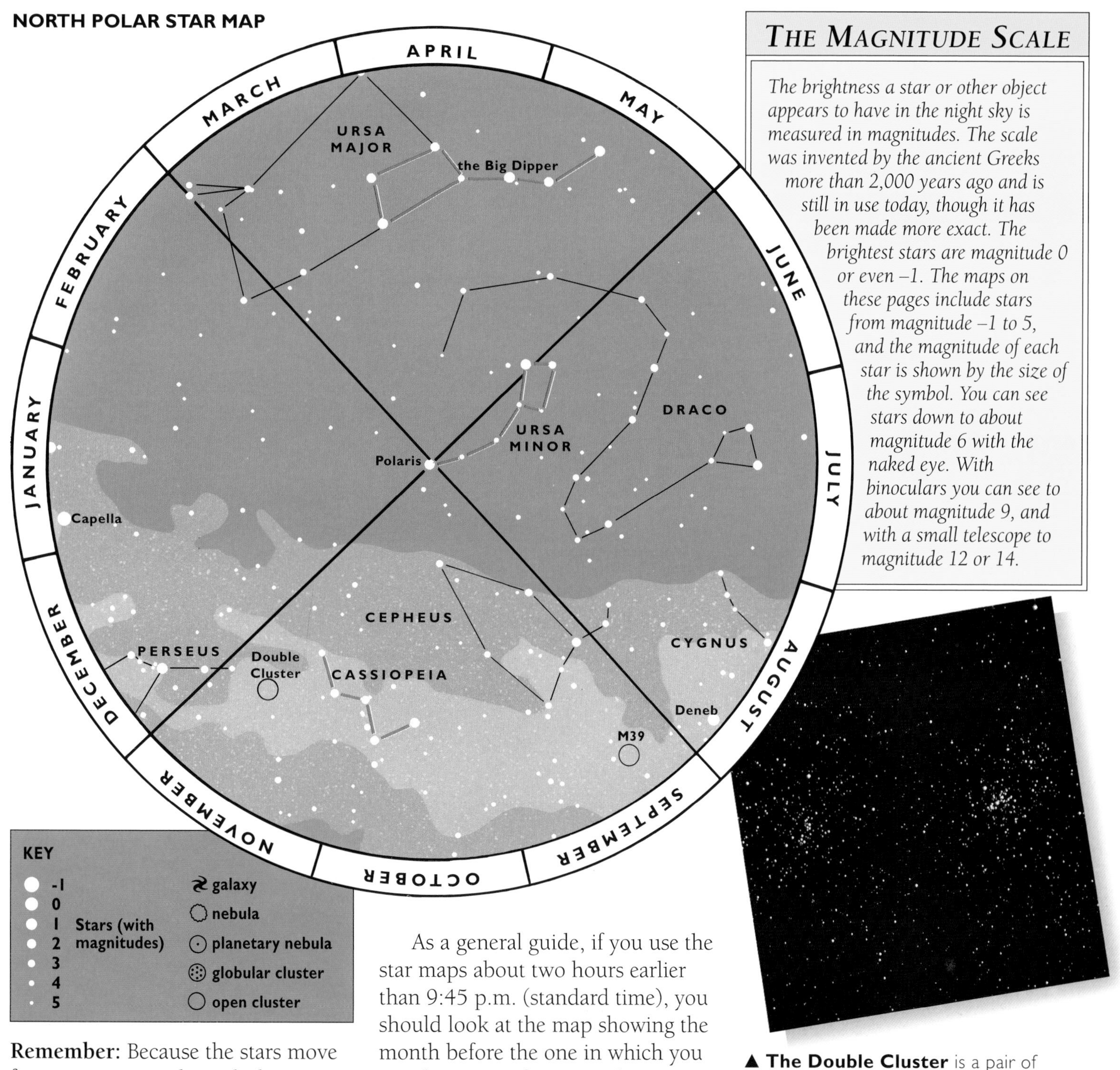

The Magnitude Scale

The brightness a star or other object appears to have in the night sky is measured in magnitudes. The scale was invented by the ancient Greeks more than 2,000 years ago and is still in use today, though it has been made more exact. The brightest stars are magnitude 0 or even –1. The maps on these pages include stars from magnitude –1 to 5, and the magnitude of each star is shown by the size of the symbol. You can see stars down to about magnitude 6 with the naked eye. With binoculars you can see to about magnitude 9, and with a small telescope to magnitude 12 or 14.

▲ **The Double Cluster** is a pair of clusters about 7,500 light-years away. It can be seen with the naked eye, but is beautiful when seen through binoculars.

Remember: Because the stars move from east to west through the year, the months are given in reverse order on the star maps.

As a general guide, if you use the star maps about two hours earlier than 9:45 p.m. (standard time), you should look at the map showing the month before the one in which you are observing. If you use the maps about two hours later, then look at the map with the following month.

Star maps

From September to November there are few bright stars because we are looking away from the main band of the Milky Way (*see pages 52–53*). The Square of Pegasus is the best pattern to use as a guide. It is larger than you might think, and it is not a perfect square.

Between June and August the brightest part of the Milky Way can be seen, stretching from Cygnus, with its Great Rift of dust clouds, down to Sagittarius and Scorpius, where the center of the Galaxy lies. The area is full of clusters and nebulae (*see pages 54–57*).

▲ **You can find the galaxy M31** in Andromeda. M31 is a spiral galaxy close to our own (*see page 58*), and it is quite similar. At 2.5 million light-years from Earth, it is the most distant object that is easily spotted. You can see it with the naked eye, but binoculars show it clearly.

KEY

-1	Stars (with magnitudes)		galaxy
0			nebula
1			planetary nebula
2			globular cluster
3			open cluster
4			
5			

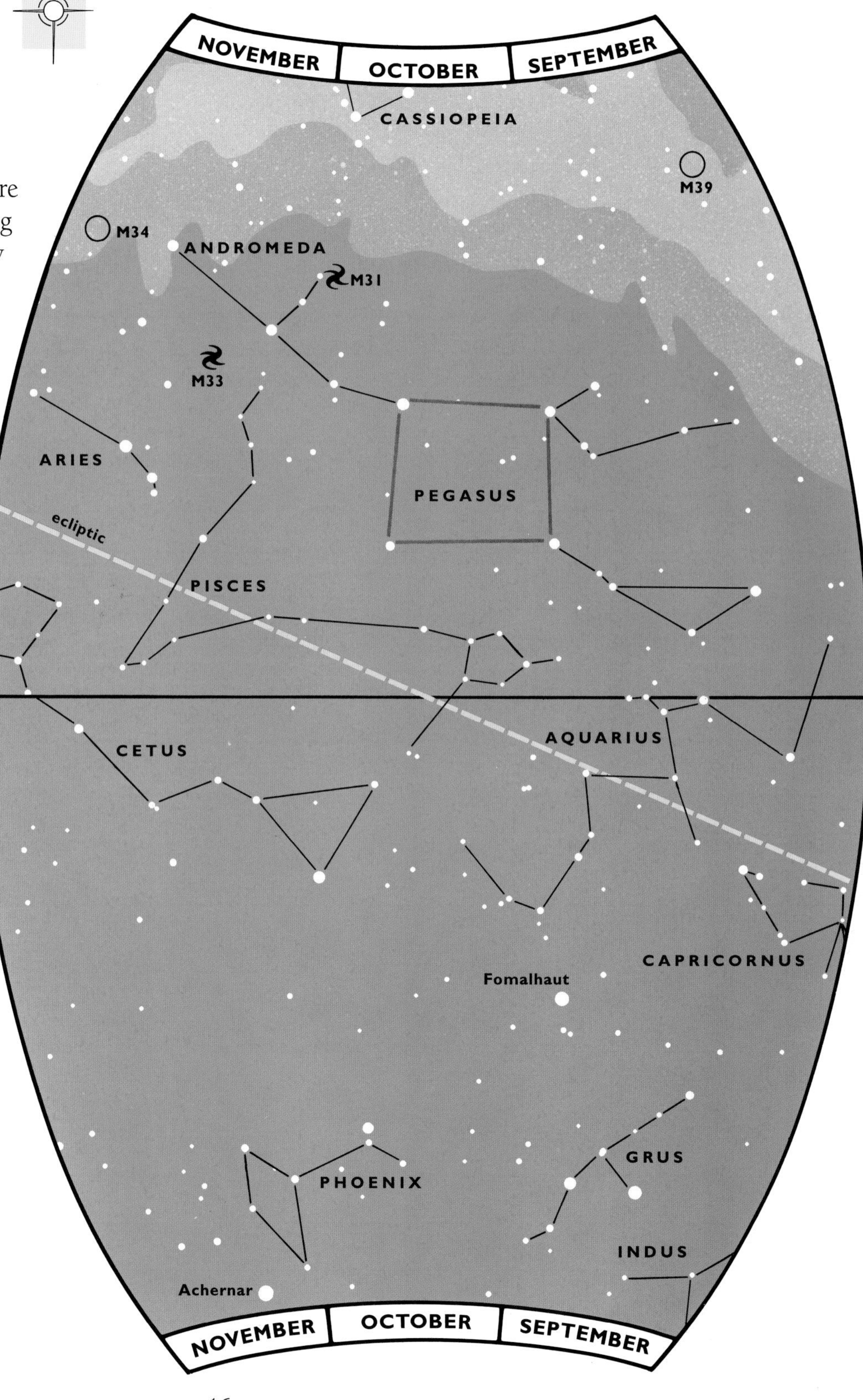

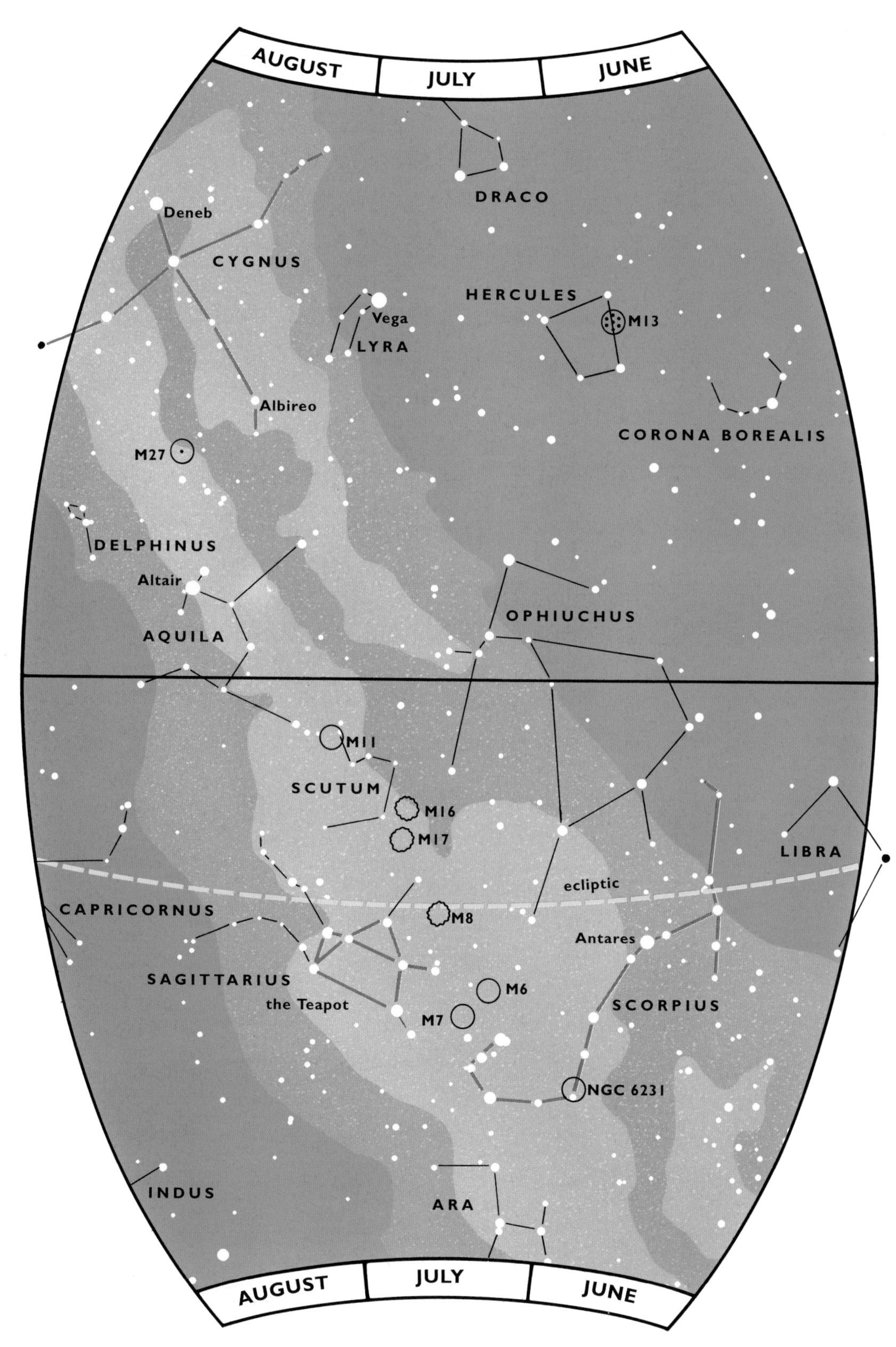

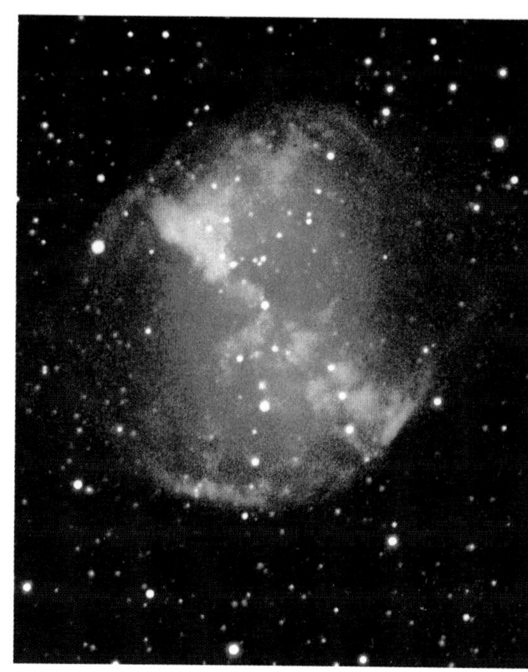

▲ **The Dumbbell Nebula,** also known as M27, is the brightest planetary nebula (*see page* 40). It can just be seen with binoculars and looks like a tiny, hazy star. The colors show up only in photographs.

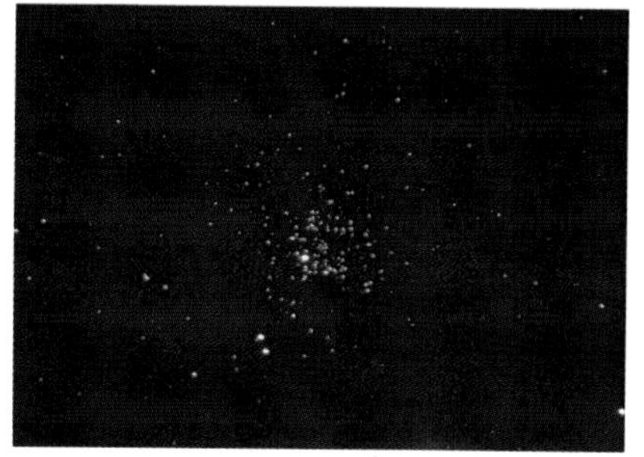

▲ **M11, an open cluster,** is also called the Wild Duck Cluster because of its V-shape of stars, which looks just like a flock of ducks in flight. The "M" numbers that are used to name some of the things we see in space come from a list of many bright clusters, nebulae, and galaxies, which was put together by the 18th-century French astronomer Charles Messier.

Star maps

Two of the most famous star patterns are in the sky between March and May: the Big Dipper, which can be seen from the northern hemisphere, and the Southern Cross, visible from the southern hemisphere. Between them lies Leo, the lion, one of the few star groups that looks like its name.

From December to February, the Milky Way is in the sky again, but it is not as bright as it is between June and August. The stars of Orion are bright and make up one of the best-known patterns in the sky. Follow the line of the three stars that form Orion's "Belt" to find, in one direction, Sirius (the brightest star) and, in the other direction, Aldebaran and the Pleiades.

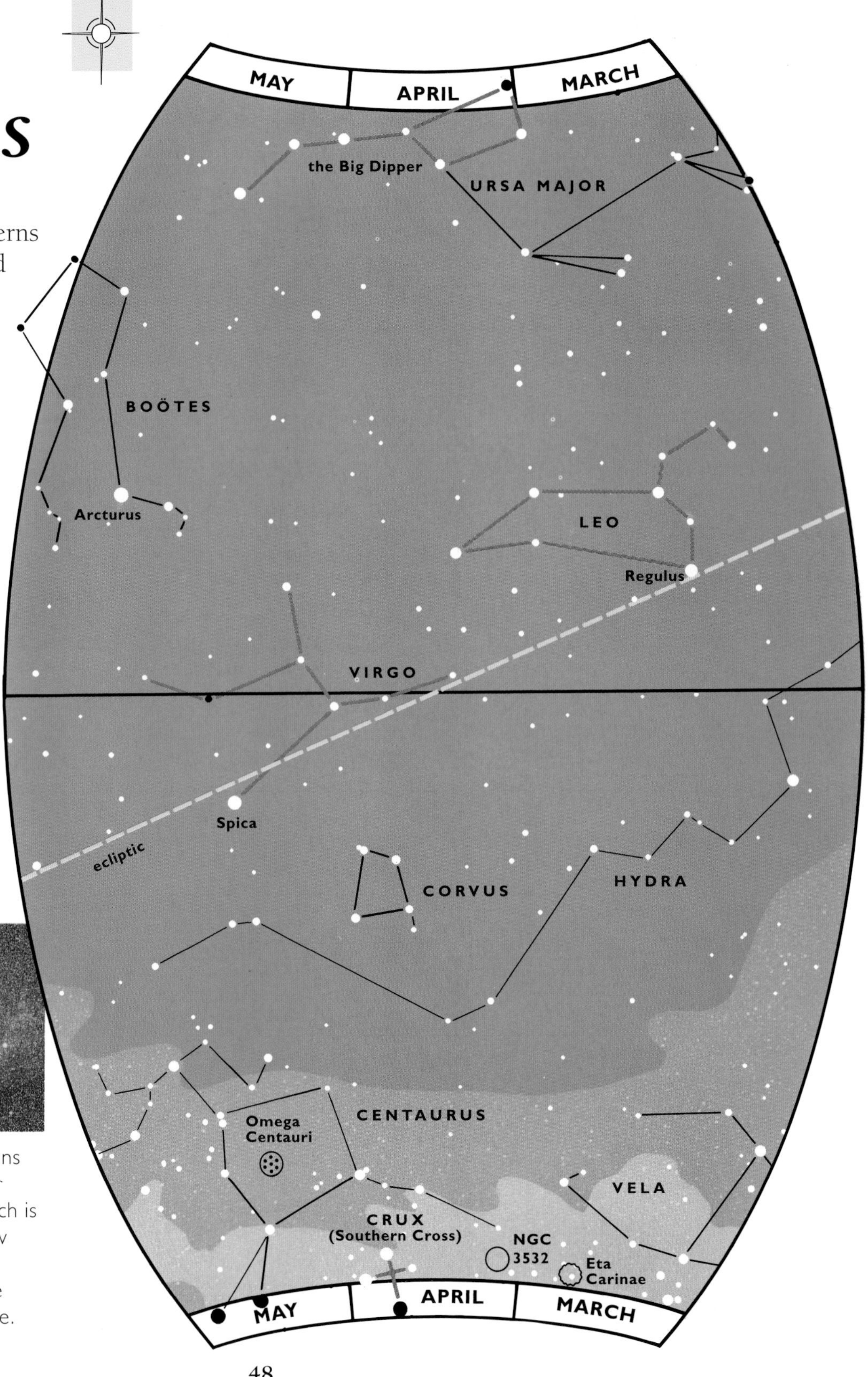

▲ **The Eta Carinae Nebula** contains some of the most massive stars in our galaxy. The star Eta Carinae itself, which is within the nebula, is unstable. It is now quite faint, but in 1843 it suddenly became almost as bright as Sirius. The nebula can be seen with the naked eye.

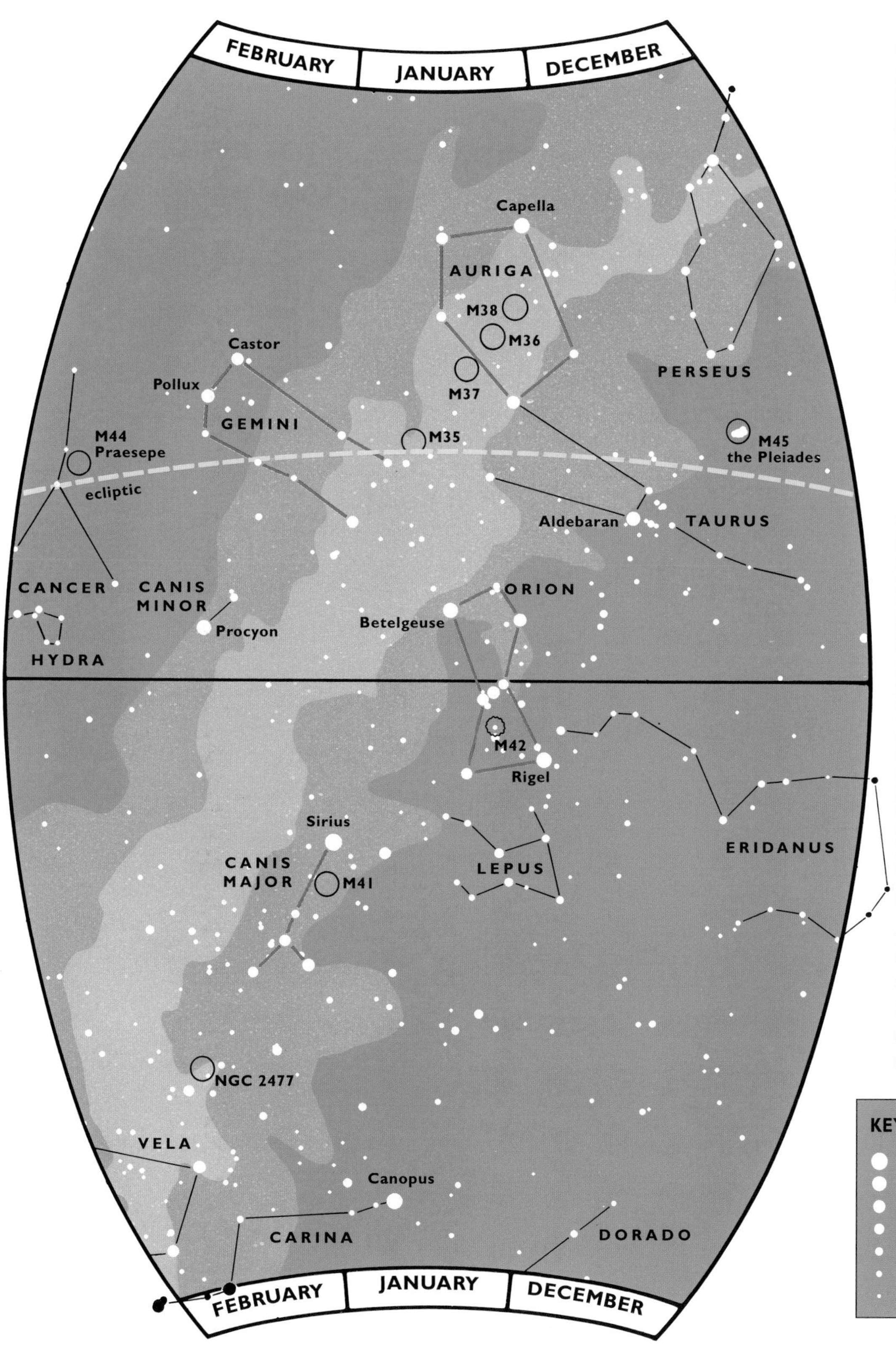

▲ **The Pleiades** is the clearest open cluster in the sky. It is also called the Seven Sisters, although many people can see more than seven stars. Try counting the stars using your eyes alone, then with binoculars—you will probably lose track!

▲ **The Orion Nebula** (M42) is the brightest nebula in the sky. On a clear, dark night it is easily seen with the naked eye. With binoculars, you can see newly born stars inside it—that is, stars born about 100,000 years ago!

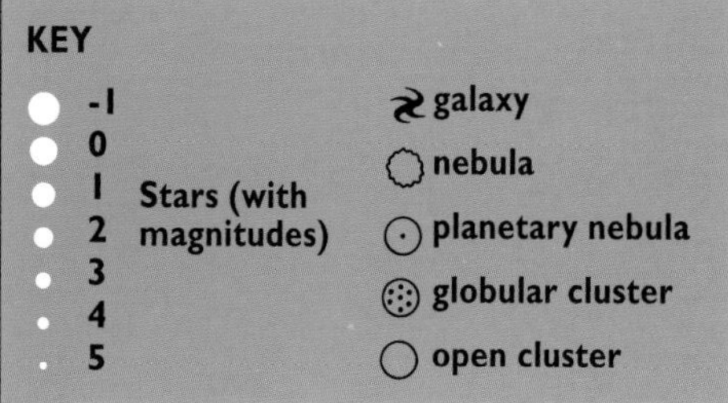

SOUTH POLAR STAR MAP

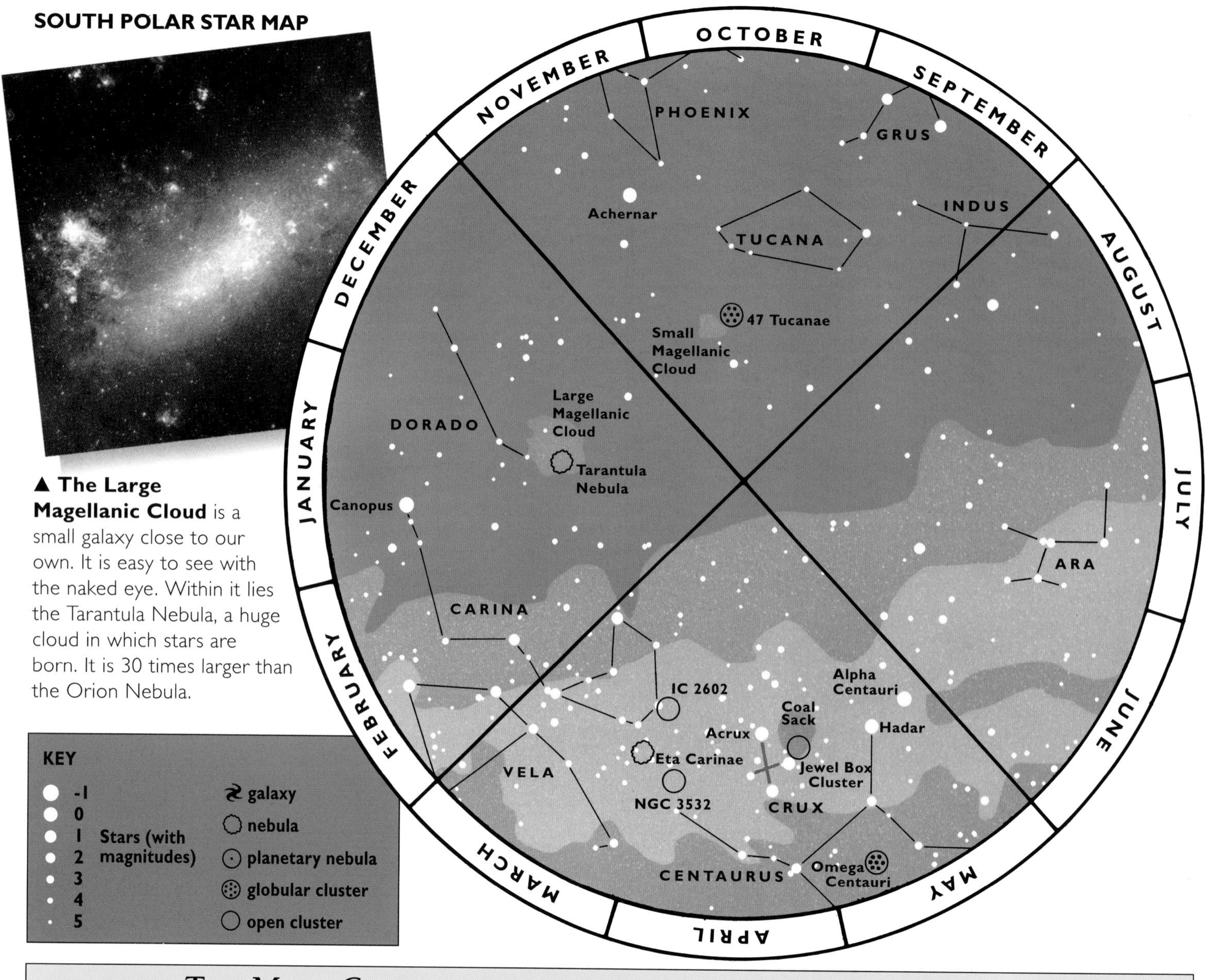

▲ **The Large Magellanic Cloud** is a small galaxy close to our own. It is easy to see with the naked eye. Within it lies the Tarantula Nebula, a huge cloud in which stars are born. It is 30 times larger than the Orion Nebula.

THE MAIN CONSTELLATIONS AND WHAT THEIR NAMES MEAN

Andromeda	Andromeda*
Aquarius	The Water-Bringer
Aquila	Eagle
Ara	Altar
Aries	Ram
Auriga	Charioteer
Boötes	Herdsman
Cancer	Crab
Canis Major	Great Dog
Canis Minor	Small Dog
Capricornus	Sea Goat
Carina	Keel
Cassiopeia	Cassiopeia*
Centaurus	Centaur
Cepheus	Cepheus*
Cetus	Whale
Corona Borealis	Northern Crown
Corvus	Crow
Crux	Cross
Cygnus	Swan
Delphinus	Dolphin
Dorado	Swordfish
Draco	Dragon
Eridanus	River Eridanus
Gemini	Twins
Grus	Crane
Hercules	Hercules*
Hydra	Water Snake
Indus	Indian
Leo	Lion
Lepus	Hare
Libra	Scales
Lyra	Lyre
Ophiuchus	Serpent-Bearer
Orion	Hunter
Pegasus	Winged Horse*
Perseus	Perseus*
Phoenix	Phoenix
Pisces	Fish
Sagittarius	Archer
Scorpius	Scorpion
Scutum	Shield
Taurus	Bull
Tucana	Toucan
Ursa Major	Great Bear
Ursa Minor	Small Bear
Vela	Sails
Virgo	Virgin

*names from ancient mythology

Patterns in the sky

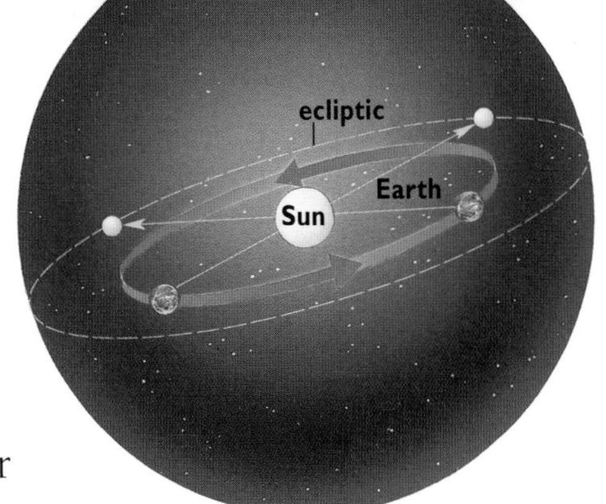

▲ **As the Earth** travels in its orbit, the Sun appears to move along a path around the sky, called the ecliptic. The ecliptic is high in the daytime sky in summer and low in winter.

Ever since our earliest ancestors gazed up at the night sky, humans have been fascinated by the stars. Many of the names we give to star patterns or constellations come from those that were used in the Near East (modern Iraq) thousands of years ago. However, different peoples around the world have their own names for the constellations.

Few of the patterns look like the things their names suggest, but that is not the point of them. At a time when information was passed from generation to generation by word of mouth rather than by writing, the stories that were told about the star patterns helped people to remember their way around the sky.

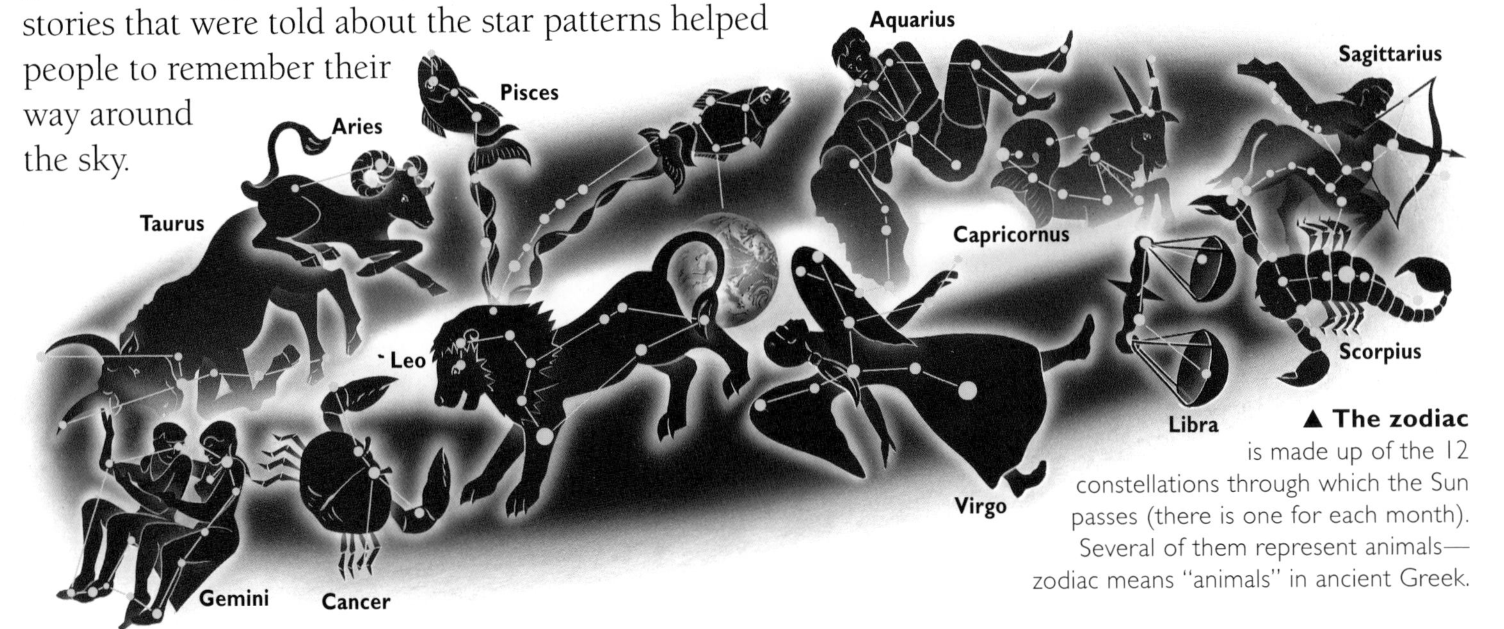

▲ **The zodiac** is made up of the 12 constellations through which the Sun passes (there is one for each month). Several of them represent animals—zodiac means "animals" in ancient Greek.

Following the Sun
The Sun's yearly path around the sky has always interested people. Although the Sun's brightness drowns out starlight, the stars are still there by day. Month by month, the Sun passes through 12 constellations, and these make up what is known as the zodiac.

The exact path of the Sun, called the ecliptic, is marked on the star maps on the previous pages. The Moon and planets also stay close to this line. So, if you see a bright star near the ecliptic but cannot find it on the maps, it is probably one of the planets Mercury, Venus, Mars, Jupiter, or Saturn.

Unlike stars, planets rarely twinkle. This is because a star is a point of light, so its beam is easily affected by the Earth's atmosphere, which makes the star appear to twinkle. Many beams of light reach Earth from a planet, so a planet's light is less affected by the atmosphere and does not flicker.

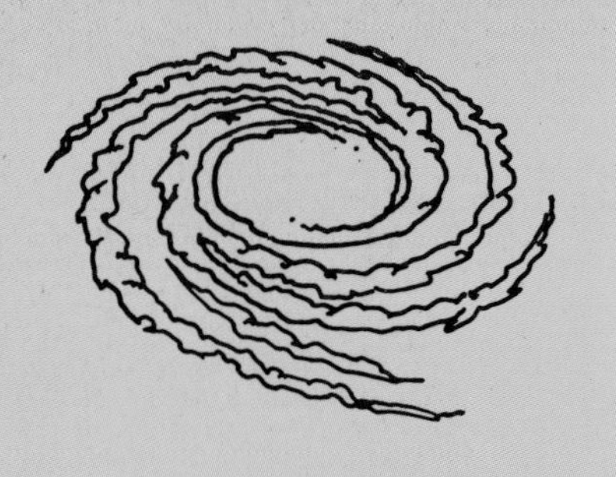

THE MILKY WAY

ON SOME DARK, CLEAR NIGHTS, WE CAN SEE A HAZY BAND OF LIGHT stretching right around the sky. This is the Milky Way: we are looking at our own huge galaxy of stars from the inside. Almost everything we can see in the night sky is part of our galaxy.

Plan of the Galaxy

Go out into the country on a clear night, using the star maps (*pages 45–50*) as a guide, and you will see the Milky Way. As its name suggests, it looks like a river of milky light running through the sky. Look at it through binoculars and you will see that it is made up of countless faint stars.

Astronomers now know that we are looking at a huge system of stars from the inside. Our sun is part of this system, which is called the Galaxy—the word "galaxy" comes from the ancient Greek name for the Milky Way. There are other similar systems outside our own, and these are also called galaxies, but we call ours the Milky Way Galaxy.

▲ **The Milky Way** is best seen in July and August, when it crosses the sky overhead. You need to be well away from city lights to see it properly.

Stars and spiral arms

The Galaxy is shaped like a giant disk or pinwheel. It is 120,000 light-years across and contains over 100 billion stars. Its most spectacular features are arms, made up of young stars and clouds of gas and dust, that spiral out from its center. Our sun lies in a spiral arm, about 30,000 light-years from the center of the Galaxy.

The milky band we see in the sky is, in fact, the line of the Galaxy's spiral arms, crowded with stars. Outside of this band we see fewer stars; they are still part of the Milky Way Galaxy, but do not lie in the Milky Way itself.

At the center of the Galaxy is a bulge, or hub. This contains reddish stars more than 10 billion years old, with almost no gas or dust. Around this is a halo of old stars, with ball-shaped clusters of stars (known as globular clusters) mixed in.

All the stars orbit around the Galaxy's center, at speeds that depend on how far out they are. Stars in our neighborhood take about 250 million years to go around once. Those farther out take longer.

▲ **Dust clouds** hide much of the Galaxy from our view, but astronomers can still map it by detecting radio and other waves coming from the material in the spiral arms. The arms are named for the constellations we see in those directions. The Sun lies in the Orion arm.

▲ **The Milky Way Galaxy** would look something like this if we were able to see it from the outside. A huge disk with a bulge at its center makes up the main part of the Galaxy, which is crowded with stars. Other stars surround the center in a ball-shaped halo. The different parts of the Galaxy contain particular types of stars, and these are shown in the pictures at right.

The spiral arms look as if they are winding up, like a blob of thick paint being stirred into a pot of another color. But they are not winding up. They move more like waves slowly rippling around the Galaxy. Stars can move in and out of the spiral arms faster than the arms themselves turn.

▲ **If we could slice through** the edge of the Galaxy, we would see that the Sun is almost exactly in the central line. The disk is about 1,300 light-years thick at this point. The center of the Galaxy is hidden from view by dust clouds. These clouds also create the dark band, called the Great Rift, which we see in the Milky Way.

▲ **In the central bulge and halo** of the Galaxy are old stars, which formed billions of years ago. In the spiral arms are younger stars, together with gas and dust from which new stars are still forming.

▲ **The Horsehead Nebula** in Orion is a swirl of dust seen against a bright red nebula.

▼ **Dust lanes** show up against a background of glowing gases. The dark dust lanes in the Trifid Nebula *(see the photograph below)* make it look as if it is divided into three—its name means "split in three" in Latin.

dust lane

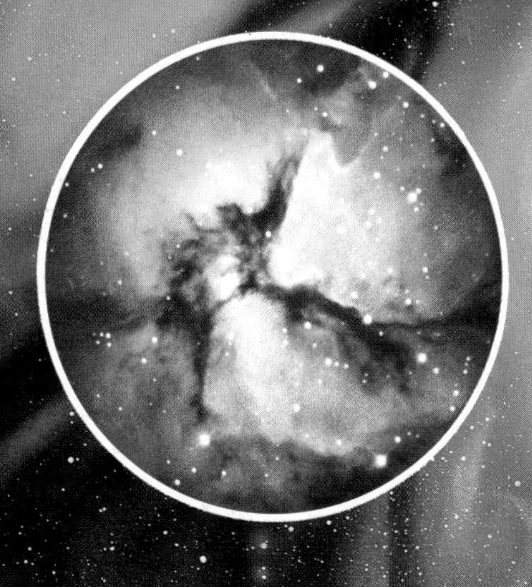

Bright and dark nebulae

Nebulae make some of the most spectacular pictures in any book on astronomy. Some people buy telescopes especially to see these dark swirls and colorful clouds for themselves. But they may be disappointed because, even with the help of a telescope, our eyes are very bad at seeing color in dim objects.

What gives nebulae their shapes and glowing colors? Nebulae are clouds of gas and dust. They are made mostly of hydrogen gas, with some other gases such as oxygen, nitrogen, and sulfur mixed in. It is combinations of these, together with dust and the light from nearby stars, that give nebulae their amazing shapes and colors.

Different combinations of gases and dust create different effects. Some of these effects are shown in the nebula that is pictured across these pages. To help understand the picture, look at the map of the nebula on the opposite page.

Bright nebulae

Unlike stars, nebulae have no light of their own. But, if there is a hot star within a few light-years, the gas in a nebula will probably shine brightly. Hot stars give out lots of ultraviolet light, which is absorbed by the gas and is given out again as particular colors. Nebulae that glow like this are known as emission nebulae. Hydrogen gas glows red, while oxygen and helium are green. The spectrum of the cloud (*see page 35*) will show bright lines of these colors, telling us which gases are present.

Dust among the stars

There is a lot of dust in space. It comes from stars, and it may be blown off them as they shine or in a supernova explosion.

Why don't our eyes see nebulae in the same colors as photographs? Next time there is a Full Moon, try looking at objects outdoors. Green trees and red doors all appear gray. This is because our eyes have two types of light-sensitive cells: rods and cones. The cones detect colors, but they need plenty of light. The rods can see in dim light, but they give only a black-and-white view. Not even the largest telescope can make a nebula bright enough for the cones in our eyes to see the color. So, we see all nebulae as gray. Because a photograph or an electronic detector, called a charge-coupled device (CCD), can go on absorbing and building up light over a period of time, they can pick up color.

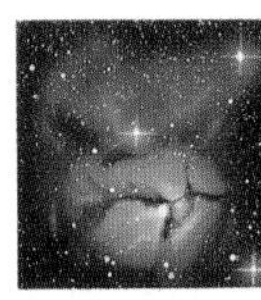

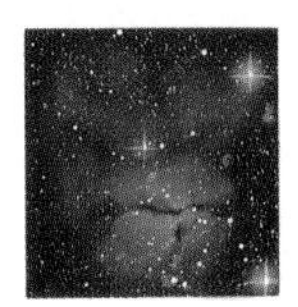

▲ **This diagram shows the spectrum of a nebula** with bright lines of color made by the different gases in it. Color film picks up these colors because it has three layers, each sensitive to blue, green, or red light. When the three layers are looked at together, the true color of the nebula can be seen.

The dust is probably made of particles of carbon or iron with a coating of ice or frozen gas. The particles are about the size of those in wood smoke. Like smoke, the dust appears bluish when light shines on it, so dust clouds look blue if there are hot stars nearby. These are known as reflection nebulae. Usually, however, the dust blots out the light from more distant stars or nebulae, and the clouds appear dark.

▶ **This is a map of the nebula** pictured in the background illustration across these pages. The map shows the nebula from above. To the right of the nebula, a nearby hot star shines on the dust, which looks blue. Stars inside the nebula make its hydrogen glow red. The bright areas look green to the eye because of a strong blue-green coloring from oxygen.

▶ **The light from a bright star** in front of the nebula is reflected by the dust as blue light. The photograph of the Rho Ophiuchi Nebula (*far right*) shows this effect.

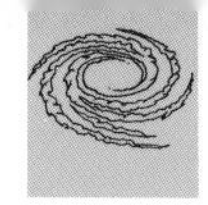

Clusters of stars

Stars usually form in groups, not on their own. Many stars will form in the same part of our Galaxy over a period of time, as a cluster. If there is only a small number of stars in the cluster, they will drift apart over a few million years and each star will make its own way through the Galaxy. If the cluster contains lots of stars, they will stay together.

There are thousands of clusters in the Milky Way, of all sizes. Some have no overall shape and are called open clusters, or galactic clusters because they all lie in the disk of the Galaxy. Others, which are found in the Galaxy's halo, are ball shaped and are called globular clusters.

▶ **M13 is a globular cluster** in the constellation of Hercules. It lies 21,000 light-years from Earth and is about 160 light-years across. A telescope is needed to see it well.

Open clusters

The members of an open cluster stay together because each star has a gravitational pull on the others. Eventually, however, members of even a large cluster can escape from the group, and as time goes by, the cluster gets looser. Most of the open clusters we see are quite young—less than 500 million years old, which is a tenth of the age of the Sun. They usually have a few hundred members, each a light-year or so from its neighbor.

Clusters can give astronomers lots of useful information because all the stars in a cluster probably formed at about the same time from the same materials and are at roughly the same distance from Earth. By looking at the types of stars in different clusters, astronomers can work out how stars age. Young clusters have many hot, blue stars that will use up their fuel within a few million years. Older clusters have no blue stars left, and their brightest stars are red giants.

Globular clusters

Surrounding the Galaxy are the globular clusters. A typical globular cluster contains about half a million stars. These are jammed into a space usually about 100 light-years across. A planet in the center of a globular cluster would be bathed in the light of hundreds of nearby stars.

The stars in a globular cluster are some of the oldest known—up to 15 billion years old. They must have formed when the Universe was young, and have stayed together all that time. There is no gas or dust in these clusters because star formation ended there billions of years ago.

There are big differences between the stars in globular and open clusters. Stars in globular clusters were born from just the light elements hydrogen and helium

▲ **Open clusters** are scattered across this part of the Milky Way near the Southern Cross. Some are groups of a few dozen stars, such as the Theta Carinae group (*below center*), while others have thousands of stars.

that were around at the beginning of the Universe. The first big stars used up their fuel quickly and exploded as supernovae (*see pages 40–41*), spreading heavier elements produced inside them through the cluster. Later stars, which are the stars that we see today, were formed with these heavier elements in them.

But there have been thousands of supernovae in the disk of the Galaxy, each spreading heavier elements. So the younger stars in open clusters have more of these elements in them than stars in globular clusters.

Looking at Clusters

You can see hundreds of clusters for yourself using just binoculars. Make sure that they are properly set for your eyes before you start observing. There are usually three things to do. First, move the eyepieces so that they are the right distance apart for you. Turn the center wheel to focus on an object using your left eye only. Finally, turn the right eyepiece to focus on the same object. Write down the settings so that you can do this easily each time you observe.

The best time to look for clusters is when the Milky Way is overhead. Lie on your back and scan along it. You will see dozens of clusters. You can find the brightest using the star maps on pages 45–50.

▶ **These binoculars** are size 7 x 50. The magnification is 7 and the lenses are 50 mm (2 inches) across—ideal for stargazing.

OTHER GALAXIES

NOT ALL THE MISTY PATCHES IN THE SKY ARE NEBULAE IN OUR GALAXY. We can look beyond the stars of the Milky Way Galaxy to see faint smudges that are other galaxies outside our own. Some of these are spirals like the Milky Way. Others have different shapes.

Types of galaxy

The Milky Way Galaxy is one of a group of about 30 galaxies known as the Local Group, so we have a few close neighbors in space. Our galaxy is one of the main members of the group, along with the Andromeda Galaxy, also known as M31, and another spiral galaxy called M33.

Spirals are not the only type of galaxy. There are ball-shaped ones, called ellipticals, but the most common type has no obvious shape and is known as irregular. Each of these types are found in the Local Group.

▲ **NGC 2997** is a typical spiral galaxy. The stars across this photograph belong to our own galaxy—we are looking outward past them.

Touring the Local Group

There are two separate neighborhoods in the Local Group, with the Milky Way Galaxy in one and the Andromeda Galaxy in the other. Not far from our own galaxy are two medium-sized irregular galaxies, called the Large and Small Magellanic Clouds.

The Magellanic Clouds and a number of smaller galaxies orbit our own galaxy. The Andromeda Galaxy also has its followers, such as the elliptical galaxy M32. At the same time, the two groups of galaxies slowly orbit each other.

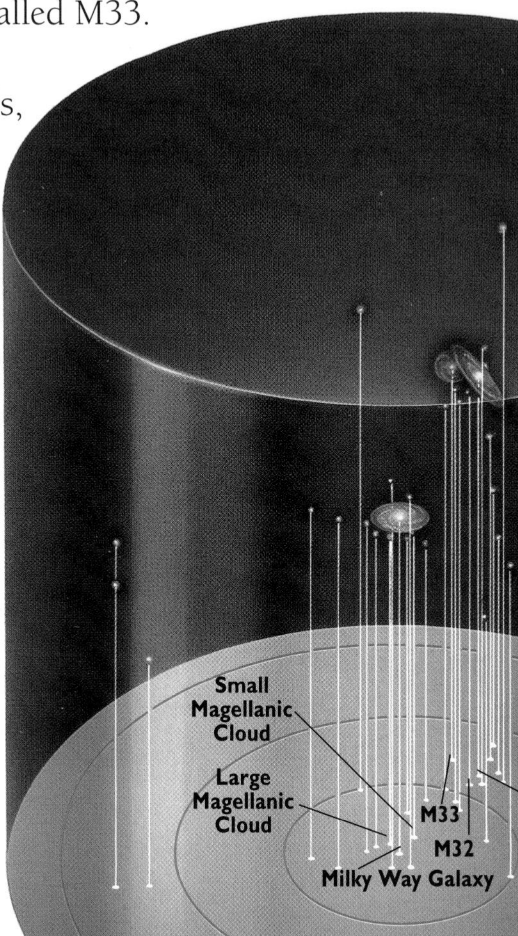

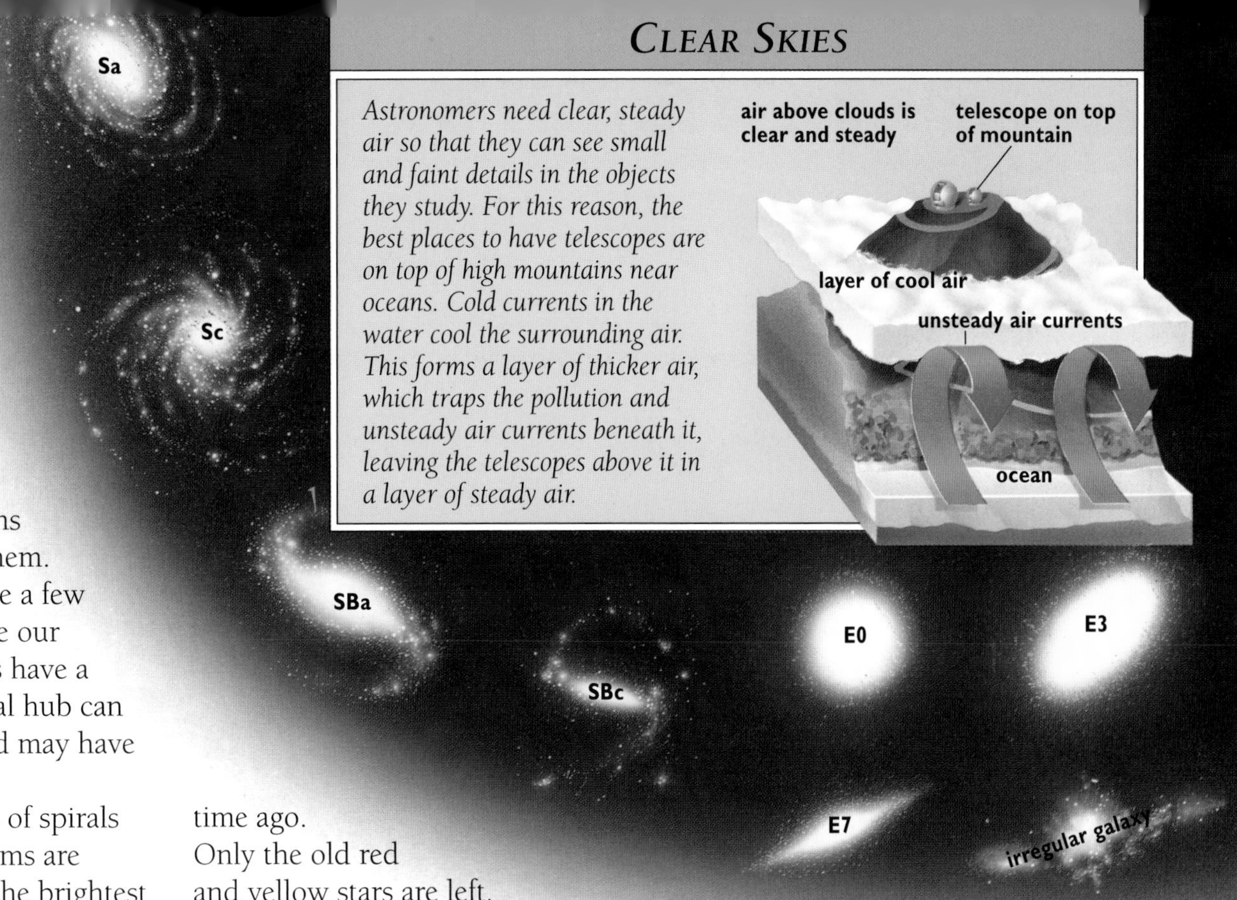

▶ **There are many** types of galaxies. Spirals and ellipticals are labeled with the letters S (spiral), SB (spiral with bar across center), or E (elliptical), and have extra letters or numbers to show their shapes.

Spiral galaxies

Much farther away, galaxies are scattered across space with millions of light-years between them. Some spiral galaxies have a few loosely wound arms, like our own galaxy, while others have a tighter shape. The central hub can be large or small, and may have a bar across it.

Color pictures of spirals show that the arms are bluish because the brightest stars in them are young O-type stars. Pink nebulae, where stars are being born, are seen along the arms. Their centers look yellowish because lots of older red giants are found there.

Cosmic footballs

Elliptical galaxies are almost always yellowish because starbirth ended there a long time ago. Only the old red and yellow stars are left. The shapes of elliptical galaxies range from circular to flattened footballs. Some are so flattened that they are lens shaped.

There are huge differences in the sizes of elliptical galaxies. Some have only a few hundred thousand stars, but giant ellipticals are the biggest galaxies known, with as many as ten trillion stars.

◀ **The Local Group** has more than two dozen galaxies within about three million light-years of space. New members are found from time to time.

Irregular galaxies

Some galaxies are just groups of stars with no obvious shape. These are the irregular galaxies and they are usually quite small. They look blue in photographs.

Weird and wonderful

Some galaxies do not fit into any group, but have clear shapes of their own. There are ring galaxies, such as the Cartwheel, and galaxies that seem to have horns, such as a pair of galaxies called the Antennae.

Usually there are signs that these galaxies have collided with others. When this happens, the stars rarely hit each other because the distances between them are so great, but gravity pulls the stars into odd orbits, making the weird shapes.

CLEAR SKIES

Astronomers need clear, steady air so that they can see small and faint details in the objects they study. For this reason, the best places to have telescopes are on top of high mountains near oceans. Cold currents in the water cool the surrounding air. This forms a layer of thicker air, which traps the pollution and unsteady air currents beneath it, leaving the telescopes above it in a layer of steady air.

▲ This is the heart of the Virgo Cluster, with the giant elliptical galaxies M84 (*right*) and M86 (*above center*). Most of the galaxies in the Virgo Cluster are spirals, but giant ellipticals lie at its center.

Clusters of galaxies

Beyond our Local Group lie other clusters of galaxies. Some of these are huge, with thousands of members. The clusters themselves are in groups, called superclusters, which are the largest known structures in the Universe.

Close to the Local Group lies the Virgo Cluster. None of its members is bright enough to be seen with the naked eye, but it covers a large area of our sky—about the size of your clenched fist held at arm's length—even though it is 50 million light-years away.

Near the heart of the Virgo Cluster lies a giant elliptical galaxy, called M87, which is the largest galaxy known. It is usual to find a giant elliptical at the center of a large galaxy cluster, even though spiral galaxies are more common.

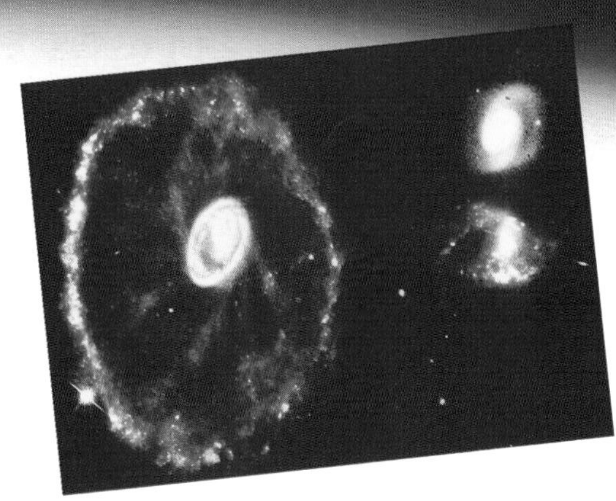

▲ The Cartwheel Galaxy (*left*) shows what can happen when two galaxies collide. A smaller galaxy passed through its center, causing a ripple of star formation to spread outward. The blue ring contains billions of new stars that have formed suddenly. Spiral arms are starting to take shape again.

Galactic cannibals

The galaxies in a large cluster are quite close together, so collisions happen from time to time. When two spiral galaxies collide, the stars might miss each other, but the gas and dust mix. The material falls to the center of the larger galaxy and is used up in a great burst of star formation. Some of the material may form a huge black hole.

So the larger galaxy eats up the smaller one, and the gas and dust

▲ Astronomers can create models of colliding galaxies on computers. This picture is based on a computer model and shows (*from left*) two spirals joining together over millions of years to make a single giant galaxy with tails of stars.

URSA MAJOR CLUSTER

VIRGO CLUSTER

CANES VENATICI CLUSTER

Local Group

◀ **The Local Supercluster** is about 100 million light-years across and contains the Local Group, the Virgo Cluster, and several other nearby clusters of galaxies. Mapping the galaxies outside the Local Supercluster (*right*) shows other superclusters, with voids of empty space between them.

from both forms into stars. The new bigger galaxy has a stronger gravitational pull, so it attracts and swallows up more galaxies. Eventually, it becomes a big galaxy, in which there is no gas and dust left, at the center of a cluster. This may be why we find giant ellipticals at the centers of clusters—they are cannibal galaxies, which have eaten up many others.

Mystery mass

Astronomers have discovered something very peculiar about the Virgo Cluster. They have found that there is more material in it than we can see.

In the same way that the orbits of double stars can tell us their masses (*see page* 38), the movements of galaxies inside a cluster can be used to work out the cluster's total mass. The brightness of the galaxies also tells us how many stars are in each one. But the Virgo Cluster seems to be about two hundred times more massive than we would expect it to be from the stars and other material we can see inside it.

There is a similar problem with almost every cluster of galaxies. No one is certain why there is this "missing mass." Some people believe that there are types of objects that we have not yet discovered. Finding this missing mass is an important task for astronomers today.

Empty space

The Virgo Cluster, the Local Group, and other galaxy clusters make up the Local Supercluster, which is about 100 million light-years across. Beyond the Local Supercluster is the Coma Supercluster, which is about 300 million light-years away.

Between these superclusters lies a "void"—a region with almost no galaxies. As astronomers have mapped the positions of galaxies across space, they have found that there are lots of these voids. It seems that galaxies are arranged like a spider's web, with clusters and superclusters forming long chains with gaps between them.

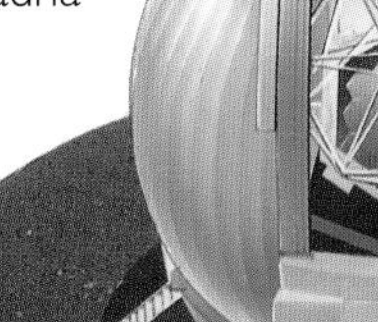

▶**The Keck Telescope** is the world's largest telescope and is used to look at very faint, distant galaxies and clusters. It has a mirror 33 feet across, and is at the top of Mauna Kea Mountain in Hawaii.

Active galaxies

There is something very unusual going on inside some galaxies. They stand out from all the others because they are much brighter or they look quite different in some other way.

Astronomers are interested in these strange galaxies because, by discovering what is going on there, they can learn more about what happens inside ordinary galaxies.

Usually, galaxies stand out because some sort of activity is seen in their centers. These are called active galaxies. Sometimes an active galaxy looks normal, except that it produces unusually strong radio waves, so it is known as a radio galaxy. Other galaxies have centers that are so bright that the galaxies themselves are lost in the glare, and these are called quasars.

▲ **From the galaxy M87,** a jet of gas streams for nearly 6,000 light-years. The very center of the galaxy is marked not by stars but by a bright spot which may contain a black hole with over a billion times the mass of the Sun.

Jets and black holes

The giant elliptical galaxy, M87, is a radio galaxy. It has a long jet of material stretching for about 6,000 light-years from its center, giving off light, radio energy, and X-rays. There are many galaxies with one or two jets like this.

At the ends of these jets are huge clouds of material, called lobes, which give off radio waves. The lobes may be bigger than the galaxy itself and are often found outside it.

A huge amount of power is needed to create these jets and lobes. The center of a typical active galaxy gives off 100 times as much energy as our own galaxy. Astronomers believe that it comes from a giant black hole.

A black hole that has formed from a single star (*see page 43*) has a mass of just a few Suns, but a black hole at the heart of an active galaxy may have a mass of a billion Suns. A stream of stars and other material falls toward the black hole, forming a hot swirling disk around it. The heat causes gas to flow off the disk in jets.

The jets stream outward through the galaxy until they meet thin clouds of gas at its edge, where they spread out to form the lobes.

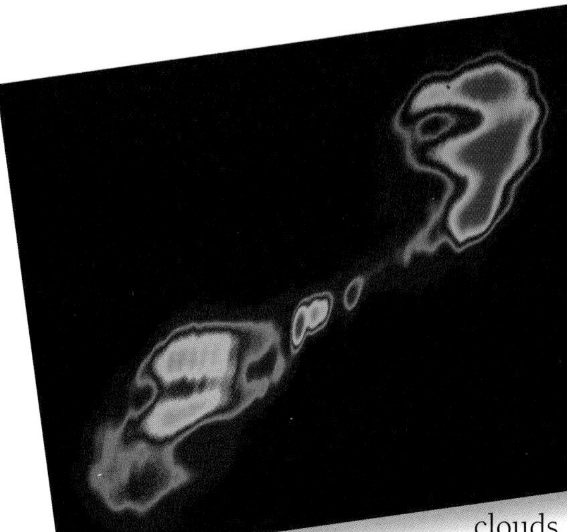

▲ **Centaurus A** is the nearest radio galaxy to Earth. Seen through binoculars, it is just a tiny blob, but a radio telescope shows huge lobes of gas on either side of it.

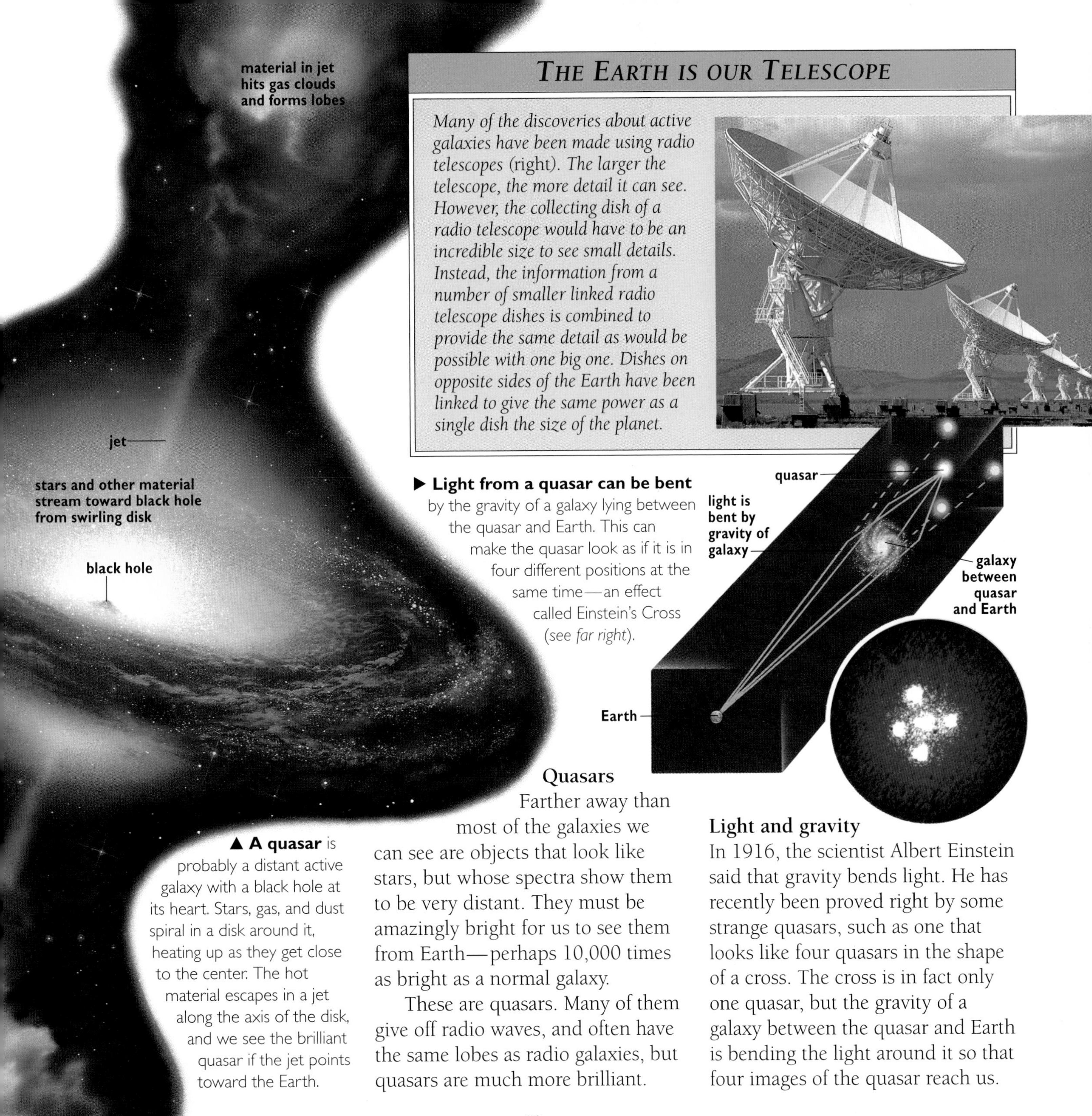

The Earth is our Telescope

Many of the discoveries about active galaxies have been made using radio telescopes (right). The larger the telescope, the more detail it can see. However, the collecting dish of a radio telescope would have to be an incredible size to see small details. Instead, the information from a number of smaller linked radio telescope dishes is combined to provide the same detail as would be possible with one big one. Dishes on opposite sides of the Earth have been linked to give the same power as a single dish the size of the planet.

▶ **Light from a quasar can be bent** by the gravity of a galaxy lying between the quasar and Earth. This can make the quasar look as if it is in four different positions at the same time—an effect called Einstein's Cross (*see far right*).

▲ **A quasar** is probably a distant active galaxy with a black hole at its heart. Stars, gas, and dust spiral in a disk around it, heating up as they get close to the center. The hot material escapes in a jet along the axis of the disk, and we see the brilliant quasar if the jet points toward the Earth.

Quasars

Farther away than most of the galaxies we can see are objects that look like stars, but whose spectra show them to be very distant. They must be amazingly bright for us to see them from Earth—perhaps 10,000 times as bright as a normal galaxy.

These are quasars. Many of them give off radio waves, and often have the same lobes as radio galaxies, but quasars are much more brilliant.

Light and gravity

In 1916, the scientist Albert Einstein said that gravity bends light. He has recently been proved right by some strange quasars, such as one that looks like four quasars in the shape of a cross. The cross is in fact only one quasar, but the gravity of a galaxy between the quasar and Earth is bending the light around it so that four images of the quasar reach us.

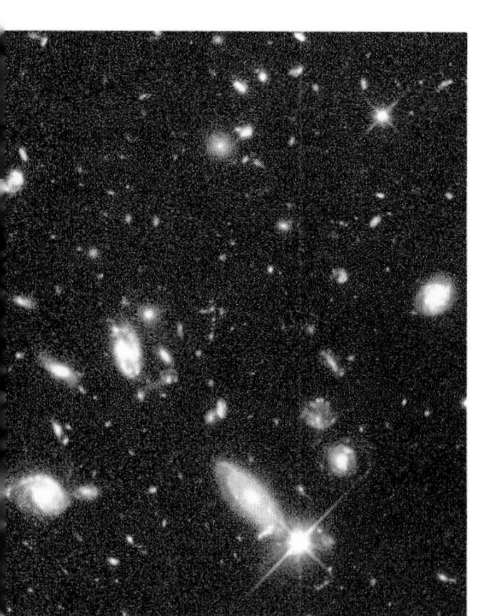

▲ **These galaxies are the** most distant ever seen by the Hubble Space Telescope. Some may have formed less than a billion years after the Big Bang, and this picture shows them early in their life.

The origin of the Universe

It might seem impossible to find out what happened at the beginning of the Universe, or even to discover when it took place. Yet astronomers believe they have a good idea about how the Universe formed. They can even work out what was going on less than a second after the Universe began.

One of the keys to this is the speed of light. Because light always travels at 186,300 miles per second, we know that the farther away we look, the longer the light has been traveling to us. So, when we look at the most distant quasars billions of light-years away, we are seeing them as they were billions of years ago.

The Doppler shifts of these distant quasars also show that they are moving away from us. The farther away they are, the faster they are moving. This is because the Universe is expanding.

The Big Bang

As the Universe expands, everything in it is moving away from everything else. Because we know the Universe is expanding, we also know that it must have been much smaller in the past, with everything much closer together. By following this expansion backward, it should be possible to determine when it began.

Astronomers believe that the Universe started expanding about 15 billion years ago. At that time, it was a very different place. Everything must have been so close together that the Universe was a tiny point, called a singularity, which is smaller even than an atom. It must also have been billions of billions of degrees hot.

Like a huge explosion, which throws out material in all directions, the Universe burst into being from this tiny point. We call it the Big Bang, and it must have been so immense that material is still rushing away.

(immediately after the Big Bang) everything is incredibly hot and expanding rapidly

(300,000 years after the Big Bang) swirling clouds of gas are formed

singularity

the Big Bang

▲ **The Big Bang** started from a singularity about 15 billion years ago. Everything—including even space and time—started at that moment, so it is impossible to talk about anything happening before the Big Bang. Space is expanding—nothing exists outside it, not even space.

◀ **The first atom** was hydrogen, the building block of the Universe. With just one proton at its center and one electron, it is the simplest atom.

(after 1 billion years) first galaxies with stars and spiral arms appear

(after several million years) protogalaxies and quasars begin to form

Within minutes of the Big Bang, the Universe had expanded to several light-years across. As it expanded, it also cooled.

The fierce temperatures at the beginning of the Universe meant that the atoms and elements that make up everything we know today (*see page 11*) could not exist. But about 300,000 years after the Big Bang, the temperature had dropped to about 5,000°F, and the first true atoms formed. Strings of hot, swirling gas clouds soon filled the Universe, all expanding rapidly.

Millions of years after the Big Bang, the first galaxies—called protogalaxies—began to form, probably with quasars at their hearts. True galaxies with stars and spiral arms appeared after one billion years.

Seeing the past

Astronomers can see something of what the early Universe looked like. The Hubble Space Telescope has looked at very distant galaxies, seeing them as they were billions of years ago, and has found that they were very different from today's galaxies. People in those distant galaxies would be seeing our galaxy as it was billions of years ago, too! Astronomers are trying to find out how these early galaxies turned into the types we know today.

The end of the Universe?

The Sun and other stars will go on shining for billions more years. Eventually, however, they will all fade and die, and the Universe will be filled with cold material.

But before this happens the Universe may start to collapse down to a single point again—in what is called the Big Crunch. This would happen if there was enough material in the Universe for the pull of its gravity to slow down and stop the expansion. It seems that there is not enough material to do this, but if there is more missing mass than we realize, this could be what happens.

Microwaves from the Sky

Everywhere in the Universe there is a faint glow. This is the remains of the energy from the Big Bang itself at about 300,000 years after the start. In 1992, small differences were found in the temperature across the glow—the hotter areas are where the gas started to form into what would become clusters of galaxies.

The glow reaches Earth in the form of microwaves—similar to those that cook food, but much less powerful. The microwaves are picked up by special equipment, such as the COBE (Cosmic Background Explorer) satellite, which first mapped the differences in temperature. Recent maps are able to show much greater detail.

COBE satellite

▲The COBE satellite map of the sky shows the glow from the Big Bang. Warmer areas are pink and red. Blue areas are cooler. The background picture to this box is a more recent map of a small area of the sky, made by ground-based equipment. Slightly warmer areas show up as white.

Glossary

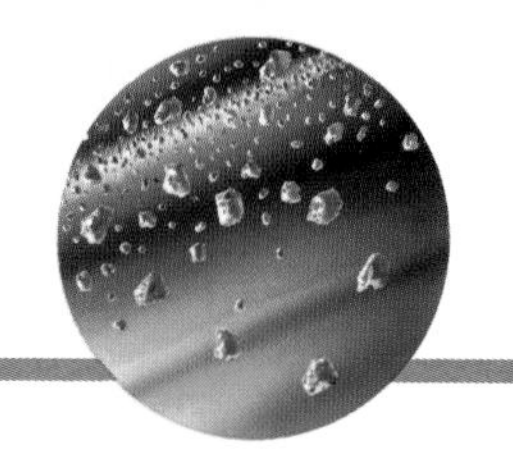

Most words to do with space are explained in the text. So try looking in the index if you cannot find the word you are looking for here.

asteroid A rocky body less than about 600 miles in diameter, which orbits the Sun usually between Mars and Jupiter.

astronaut A person who travels in space. Russian astronauts are called cosmonauts.

astronomer A person who studies **astronomy**, which is the study of the stars and other objects in the sky.

atmosphere Gases that surround a planet, held there by the pull of the planet's gravity.

atom A tiny particle that makes up all materials. Atoms are made of even smaller particles called **protons**, **neutrons**, and **electrons**. Each type of atom has a particular number of these particles. The protons and neutrons are found at the center of the atom, in its nucleus, while the electrons move around outside the nucleus.

axis The imaginary line through a spinning object around which the object rotates.

billion One thousand million.

black hole A tiny point in space into which a vast amount of material is crammed. The pull of its gravity is so strong that nothing can escape from it, including light.

carbon An element, which is found in all forms of life on Earth.

comet A piece of ice and rock a few miles across, which gives off long tails of gas and dust when it comes close to the Sun.

constellation A particular grouping or pattern of stars in the sky.

crater A bowl-shaped dip in the surface of a planet or moon, usually caused by the impact of an object from space. It has a raised rim around its edge.

electron (*see* **atom**)

element A particular arrangement of protons, neutrons, and electrons in an atom. There are over 100 different elements, including hydrogen, carbon, and iron.

ellipse A flattened circle. All orbits are in the shape of ellipses, although some are nearly circular.

energy The ability of something to do work, such as cause movement or increase temperature. Heat and light are forms of energy.

equator An imaginary line around the Earth, halfway between the north and south poles.

galaxy A collection of hundreds of billions of stars and other material such as gas and dust, all held together by gravity.

gas Material in which particles are widely separated and move around freely at high speeds.

geyser A jet of hot gas or water that shoots out of a crack, usually in rock.

gravity A force that pulls every object in the Universe toward every other object. The more massive an object is, the stronger is the pull of its gravity.

greenhouse On Earth, a glass building used for growing plants because it traps the Sun's heat and builds up warmth inside it. The word is also used to describe similar warming effects, such as when gases hold and build up heat beneath them.

helium The second simplest atom after hydrogen, with two protons, two neutrons, and two electrons. It is rare on Earth, but is very common in the rest of the Universe.

hemisphere Half of a globe. Everywhere north of Earth's equator is in the northern hemisphere, and everywhere south of the equator is in the southern hemisphere.

hydrogen The simplest atom of matter, containing one proton and one electron. Hydrogen is the most common element in the Universe.

lava Liquid rock from a volcano, which turns solid when it cools.

light A type of energy, usually produced by very hot bodies. It always travels at a speed of 186,300 miles per second.

light-year The distance light travels in a year. It is six trillion (million million) miles.

liquid Material in which particles are close together, but can move around slowly.

magnetic field The area around a magnet where the pull of the magnet is felt.

magnetic pole One of two areas of a magnet where its effect is strongest.

magnification The increase in size of an object as it appears when seen through a telescope or binoculars.

mare (Pronounced mah-ray. Plural is **maria**, pronounced mah-rear.) The Latin word for "sea," which is given to the dark areas on the Moon, once thought to be seas.

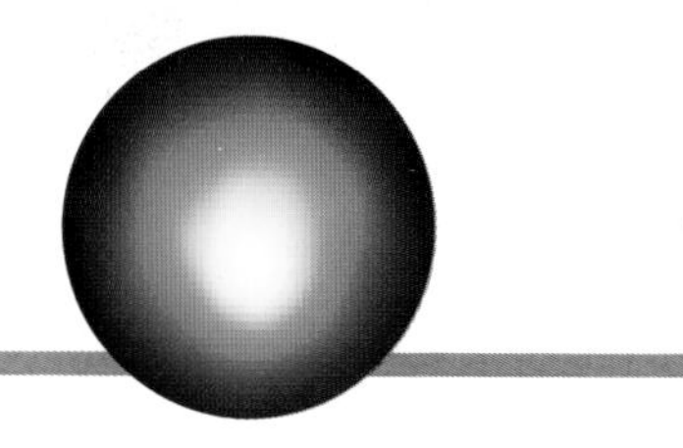

mass The amount of material in an object.

moon A natural satellite of a planet. The Moon (with a capital M) is Earth's natural satellite.

naked eye The eye alone, without the use of a telescope or binoculars.

nebula (Plural is **nebulae**, pronounced neb-you-lee.) A cloud of gas and dust in space.

neutron (*see* **atom**)

nitrogen A fairly common element in the solar system. It is the main gas in Earth's atmosphere.

nuclear energy Energy produced when the nucleus of a hydrogen atom collides with another at high speed.

nucleus The central part of an object, such as the central group of particles in an atom.

orbit The path of an object around another larger object in space. The smaller object is held in its orbit by the pull of the larger object's gravity.

oxygen An element, which is the gas we need to breathe on Earth.

particle A tiny part of something.

phase The change in the shape of the sunlit part of a planet or moon.

planet A large body that orbits a star. A planet does not shine by producing its own light, but reflects the light of the star. The Earth is a planet which orbits the Sun.

pole One end of an axis.

probe (*see* **space probe**)

proton (*see* **atom**)

quasar A very distant galaxy with a center that shines much more brightly than the rest of the galaxy.

radiation The energy sent out by an object, such as beams of light or X-rays from the Sun.

radio waves A type of radiation, similar to light, although invisible and less powerful.

satellite A body that moves around another, usually a planet. It can be a natural satellite, such as a moon, or an artificial object, such as a spacecraft.

solar system Everything that orbits the Sun or travels with it through space, such as planets, satellites, and comets.

solid Material in which particles are locked together in a fixed shape.

space The region outside the Earth which contains all other bodies in the Universe.

spacecraft A vehicle that travels through space, with or without humans inside it.

space probe An unmanned spacecraft carrying equipment to find out about space and objects within it.

spectral lines (*see* **spectrum**)

spectrum (Plural is **spectra**.) The band of colors that is produced when light is split into the individual colors of which it is made. The spectrum of light coming from an object in space can tell us a lot about the object. Lines in the spectrum, called **spectral lines**, can tell us what the object contains.

star An immense ball of gas, which produces vast amounts of heat and light.

sulfur An element often produced by volcanoes. It is used in making gunpowder.

supernova (Plural is **supernovae**, pronounced super-no-vee.) The explosion of a large star at the end of its life.

true brightness The brightness with which a star shines in space—as if it was seen from nearby. This is not the same as the brightness it appears to have when seen from distant Earth. Astronomers call true brightness "absolute brightness."

twilight The time of day just after sunset or before sunrise, while there is still light in the sky.

ultraviolet light A type of light, which comes from very hot objects. It is invisible to the eye.

Universe Everything that exists.

X-ray A type of radiation which is invisible to the eye and much more powerful than ordinary light.

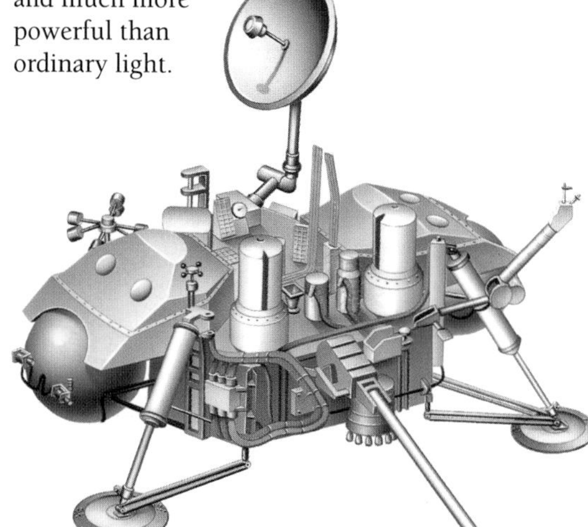

Index

Note: Page numbers in **bold** give main references; page numbers in *italic* refer to information given only in boxes or in picture captions or labels.

PICTURE ACKNOWLEDGMENTS

*All artwork by **Mainline Design**, with the exception of:*
***Bill Donohoe**, 17 (top);*
***Mark Franklin**, pages 32–33, 44–50, 61 (top);*
***Lee Gibbons** (The Wildlife Art Agency), pages 52–53;*
***Rob Jakeway**, pages 8–9, 10–11 (center), 18, 22–23 (bottom), 24–25 (bottom), 26 (bottom), 31 (bottom), 33 (bottom), 62–3 (center);*
***Colin Salmon**, 43 (right).*

PHOTOGRAPHS

l = left; r = right; t = top; c = centre; b = bottom

Endpapers: NASA; 4 NASA/Science Photo Library; 10 NASA; 12l NOAO/Science Photo Library; 12cl NASA/Science Photo Library; 12cr JISAS/Lockheed/Science Photo Library; 12r John Sanford/Science Photo Library; 14 Hermann Eisenbeiss/ Science Photo Library; 14/15 Apollo 16 Principal Investigator Frederick J. Doyle/NSSDC; 15 John Sanford/Science Photo Library; 16l NASA/Science Photo Library; 16r NASA; 17 NASA; 18 Pekka Parviainen/Science Photo Library; 19 Bill Bachman/ Science Photo Library; 20t Space Telescope Science Institute/ NASA/ Science Photo Library; 20b Rev. Ronald Rover/Science Photo Library; 21 European Space Agency/ Science Photo Library; 22/23 NASA; 24t Galaxy Picture Library; 24b NASA, 24/25 US Geological Survey/Science Photo Library; 26t Galaxy Picture Library; 26tc NASA; 26c NASA;26bc NSSDC; 26b NASA; 28 NASA; 29t John Barlow; 29b Meade Instruments Corporation; 30 NASA; 31t NASA/Science PhotoLibrary; 31b Voyager 2 Experimental team leader Dr. Bradford A. Smith NSSPC through World Data Center A for Rockets & Satellites; 34 Anglo-Australian Telescope Board; 38/39 NASA; 40l Galaxy Picture Library, 40r NASA; 41 NASA; 42 Robert Harding Picture Library; 45 NOAO/Science Photo Library; 46 NASA; 47t Kim Gordon/Science Photo Library; 47b John Sanford/Science Photo Library; 48 ROE/Anglo-Australian Telescope Board; 49 David Malin/Royal Observatory Edinburgh; 50 David Malin/Royal Observatory Edinburgh/Anglo-Australian Telescope Board; 52 NASA; 54t Anglo-Australian Telescope; 54b NASA; 55 Anglo-Australian Telescope; 56 NASA; 56/57 Luke Dodd/ Science Photo Library; 57 John Barlow; 58 David Malin/Anglo-Australian Observatory; 60t NOAO/Science Photo Library; 60b NASA; 62t Science Photo Library; 62b NRAO/AUI; 63t David Parker/Science Photo Library; 63b NASA; 64 NASA; 65 Mullard Radio Astronomy Observatory; 65 inset NASA; 67 NASA

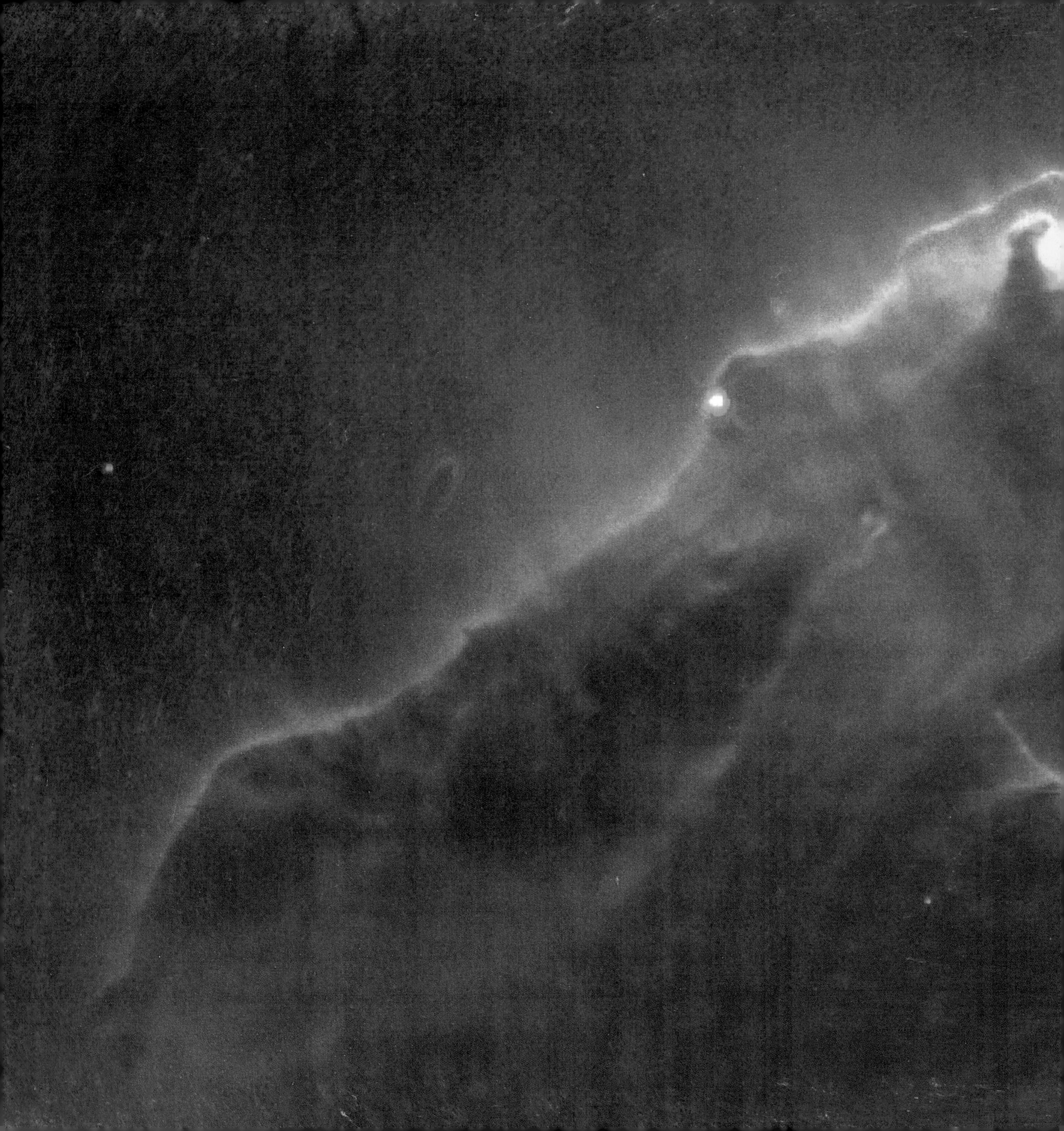